Healable Polymer Systems

RSC Polymer Chemistry Series

Series Editors:
Professor Ben Zhong Tang (Editor-in-Chief), *The Hong Kong University of Science and Technology, Hong Kong, China*
Professor Alaa S. Abd-El-Aziz, *University of Prince Edward Island, Canada*
Professor Stephen L. Craig, *Duke University, USA*
Professor Jianhua Dong, *National Natural Science Foundation of China, China*
Professor Toshio Masuda, *Fukui University of Technology, Japan*
Professor Christoph Weder, *University of Fribourg, Switzerland*

Titles in the Series:
1: Renewable Resources for Functional Polymers and Biomaterials
2: Molecular Design and Applications of Photofunctional Polymers and Materials
3: Functional Polymers for Nanomedicine
4: Fundamentals of Controlled/Living Radical Polymerization
5: Healable Polymer Systems

How to obtain future titles on publication:
A standing order plan is available for this series. A standing order will bring delivery of each new volume immediately on publication.

For further information please contact:
Book Sales Department, Royal Society of Chemistry, Thomas Graham House, Science Park, Milton Road, Cambridge, CB4 0WF, UK
Telephone: +44 (0)1223 420066, Fax: +44 (0)1223 420247
Email: booksales@rsc.org
Visit our website at www.rsc.org/books

Healable Polymer Systems

Edited by

Wayne Hayes and Barnaby W Greenland
Department of Chemistry, University of Reading, Reading, UK
Email: w.c.hayes@reading.ac.uk; b.w.greenland@reading.ac.uk

RSCPublishing

RSC Polymer Chemistry Series No. 5

ISBN: 978-1-84973-626-8
ISSN: 2044-0790

A catalogue record for this book is available from the British Library

Published by The Royal Society of Chemistry,
Thomas Graham House, Science Park, Milton Road,
Cambridge CB4 0WF, UK

Registered Charity Number 207890

For further information see our web site at www.rsc.org

Printed in the United Kingdom by Henry Ling Limited, Dorchester, DT1 1HD, UK

Preface

It is without question that polymer chemistry has made a tremendous impact upon society in the last century and the way in which day-to-day duties are carried out. After only approximately 100 years, polymer chemists have refined their art and learnt how to manipulate reactions and fabrication techniques to be able to make materials in a highly designed and predictive fashion. Reliable polymer chemistries now enable everyday items that are taken for granted, ranging from simple polyethylene carrier bags to complex block copolymers used in the production of silicon microchips, to be made to specification and on large scale. However, mankind finds itself in the 21st century in an environment that does not have endless petroleum feedstocks/energy supplies with which to continue synthesising new organic-based materials and thus there is a need to examine routes to recyclable materials that can be re-healed/mended in order to re-attain the desired function of the component or device.

The primary objective of this monograph is to provide a reference text on healable polymers that have been reported to date and cover these developments from a chemist's perspective. Healable materials span a wide range of technologies (i.e. inorganic-based materials as well as organic polymers) – this particular review focuses upon the development of organic healable polymer systems. The monograph is divided into four specialist chapters that follow on from a general introduction to the field. The introductory chapter also provides, for chemists who do not have an engineering background, an overview of the key material assessments methods that are used to study healable polymeric materials in order that someone new to the field can start to appreciate and put into perspective the physical parameters thus described. In this respect, we envisage that this monograph will be suited to a wide readership base and will be a significant asset to new researchers to this field. Each of the specialist chapters has been written by leaders in their respective fields and these

RSC Polymer Chemistry Series No. 5
Healable Polymer Systems
Edited by Wayne Hayes and Barnaby W Greenland
© The Royal Society of Chemistry 2013
Published by the Royal Society of Chemistry, www.rsc.org

reviews serve to provide a critical assessment of the healable polymer systems that have been reported to date:–

Encapsulated monomer approaches to Healable Polymers and Composites – this chapter provides a comprehensive review of the encapsulated monomer approach to healable polymer composites and also summarizes mechanistic studies of key examples from the literature;

Reversible covalent bond formation as Strategies for Healable Polymers – highly efficient reversible covalent bond formation processes such as Diels-Alder cycloadditions or thiol-ene reactions have afforded successfully healable polymer and chemically adaptive networks. This chapter assesses the reversible covalent bond formation processes that have used to this end.

Healable Supramolecular Polymeric Materials – this chapter reviews the use of weak non-covalent interactions such as electrostatic forces of attraction, hydrogen bonding, metal-ligand binding and π-π aromatic stacking in order to assemble healable polymeric networks that are constructed from relatively low molecular weight oligomeric components.

Thermodynamics of Self-Healing Reactions and their Application in Polymeric Materials – this review considers the thermodynamic parameters needed to achieve successful healable polymers. Entropic and enthalpic contributions to the Gibbs free energy during self-healing events are evaluated in the context of the chemical reactions within the material that contribute to effective healing processes.

We are extremely grateful to the authors of the four specialist chapters, it has been a pleasure to work with them in the preparation of this monograph. Finally, we would both like to thank our families (Kerry and Louisa, Emma and William, respectively) for their considerable patience and understanding for the time that we have spent away from domestic duties during the assembly and editing of the chapters for this detailed review.

Wayne Hayes and Barnaby W Greenland

Contents

RSC Polymer Chemistry Series No. 5
Healable Polymer Systems
Edited by Wayne Hayes and Barnaby W Greenland
© The Royal Society of Chemistry 2013
Published by the Royal Society of Chemistry, www.rsc.org

Chapter 3   Reversible Covalent Bond Formation as a Strategy for Healable Polymer Networks   62
Christopher J. Kloxin

Chapter 4   Healable Supramolecular Polymeric Materials   92
Barnaby W. Greenland, Gina L. Fiore, Stuart J. Rowan and Christoph Weder

CHAPTER 1

Healable Polymeric Materials

BARNABY W. GREENLAND* AND WAYNE HAYES*

Department of Chemistry, The University of Reading, Whiteknights, Reading, RG6 6AD, UK
*Email: w.c.hayes@reading.ac.uk; b.w.greenland@reading.ac.uk

1.1 Introduction

Progress in technology which serves to improve living standards and increase life expectancy is frequently linked to the materials that we humans have learnt to master and manipulate. The connection between the basic fabrics that tools can be made from and the progression in human development has become so intertwined that these advances have come to define specific eras: the stone age, bronze age and iron age. For the past 40 years or so, the role that silicon has played in advancing man's ability to address significant challenges (perhaps, most notably the lunar landings from the 1960's and 1970's) *via* computer technologies cannot be underestimated. However, there is also a growing consensus that the current period may come to be known at the *Plastic Age*.[1] From the seminal discoveries of Staudinger[2] and Carothers[3] in polymer science during the 1930's, carbon and inorganic-based polymeric products have proliferated through the modern world, finding applications in all areas from inexpensive disposable packing materials to life enhancing hip replacements and life saving body armour.

In 2010, sales of raw polymeric materials topped €117Bn in the European Union, for the first time equalling the value of petrochemicals sold in the region.[4] With demand for polymeric products growing even whilst the cost of the crude oil rises, there is a clear need to move away from the culture of disposable products that society has become accustomed to. This can be

RSC Polymer Chemistry Series No. 5
Healable Polymer Systems
Edited by Wayne Hayes and Barnaby W Greenland
© The Royal Society of Chemistry 2013
Published by the Royal Society of Chemistry, www.rsc.org

achieved one of several ways: producing more polymers from renewable feedstocks; increasing the proportion of recycled polymers in circulation and increasing the lifespan of the polymeric products. It is in the latter two potential solutions that healable materials have most to offer.[5–14] As shall be discussed in Chapters 3 and 4, the reversible nature of both covalent and supramolecular bonds is frequently exploited in producing healable materials that lend themselves to efficient recyclable materials. Producing materials that can heal either small cracks or major fractures will have a significant impact on the longevity of a host of polymeric products, from sunglasses to aeroplanes.

The diverse nature of applications for polymeric materials has occurred because of the chemist's increasing ability to design and synthesise new monomers and polymeric architectures (for example: multi-block co-polymers, branched or network materials), delivering products with useful functionality and physical properties. As synthetic techniques have progressed, the materials scientist's 'toolset' for characterising the materials at the micrometre, nanometre and angstrom scales has also improved, so the ability to successfully predict and measure the properties of new polymeric materials improves year-on-year. This structure–property interplay serves to increase the speed at which innovative materials with step-changing properties can be conceived, produced and brought to the market, further enhancing the modern world.

In the formative years of the field of polymer science, in the late 1930's and early 1940's, the primary driving force of the research carried out was to produce new materials whose properties (*i.e.* strength and thermal stability) were suitable for producing inexpensive items that could be mass produced, *i.e.* polystyrene cups, PET water bottles, strong fibres for rope and vulcanised rubber for car tyres. Recently, however, research has focused on producing high value items whose properties dictate the requirement for new polymers in order to fulfill an ever expanding number of roles, for example: shape memory materials, photo-conducting or luminescence devices and composite polymers that exhibit strength surpassing that of the strongest metal alloys. By their very nature, such polymeric materials frequently necessitate a multidisciplinary approach to their study, requiring close collaboration between the synthetic chemist—who can generate and manipulate the materials that differ at the atomistic level—and the engineer, who builds the device to test the product (*i.e.* a working solar cell or printed circuit board). Optimisation of these complex systems necessitates multiple iterations, taking many man years of effort before finally settling upon a suitable balance between the time, manpower and synthetic and fabrication costs required to attain the desired design criteria.

This interdisciplinary work ethos is clearly applicable to the field of healable polymer research. The most basic definition of a healable material requires a polymer with a given strength to be damaged, reducing its physical properties, and then, at a later point, to have regained some of the lost strength. Any research plan will require at least:

   i) the design, synthesis and characterisation of new polymers;
  ii) a method to fabricate a sample suitable for mechanical testing;

iii) a mechanical test to assess the pristine sample and healing nature of the material and

iv) an iterative development cycle whereby data from the mechanical testing will feed back into the production of the next generation of materials.

In order to complete the iterative development loop successfully, from a chemist's perspective, it is important that whilst the chemistry of the new material is fully understood, it is also imperative to have an understanding both of the mechanical tests required to demonstrate healing and of the terminology and conventions used by engineers. After summarising the potential benefits of producing commercially viable healable polymers, this chapter will define the requirements and different categories of healable materials, before providing an overview, aimed at the practising chemist, concerning the techniques currently employed to study the healablity of these fascinating new materials.

1.2 Healable Polymers – Potential Applications

While producing healable materials is an intuitively rewarding challenge, it is worth considering why developing healable materials may be beneficial in differing circumstances. As noted previously, polymeric materials are now used across a wide range of applications and fulfill a multitude of roles. In many cases, the polymer itself may play little part in adding to the strength of the product but may simply provide an attractive finish, such as the outer surface coating on a modern car. In these circumstances, small scratches caused by wear and tear are unlikely to alter the safety of the passengers. However, coatings that can regain their original luster after damage are of consider- able commercial importance both from an aesthetic viewpoint and due to improved corrosion resistance. Indeed healable coatings have been put into production in recent years by manufacturers such as Nippon Paint[15] and Bayer.[16]

Focusing on higher strength materials, one can envisage the benefit of healable gas and water pipes. In these instances, producing new materials that can autonomously fix fractures *in situ*, without the need to dig up the roads, would be a clear advantage to consumers and suppliers alike. There are also more high profile (although currently lower volume) applications associated with the high strength polymers and composites that are used in performance cars, military aircraft and the current generation of passenger aeroplanes. For example, over 20% of the airframe of the new Airbus A380 'Superjumbo' is polymer composite in nature. In these situations, where safety is an overriding design criterion, it might be expected that producing a polymer that can completely regain its strength after damage would be highly desirable. The ability of a material to self-heal in an autonomous manner is especially important in composite materials, where detecting damage within the interior of these structures is inherently difficult (and less well developed) than the identification of damage sites in conventional aircrafts made from metal–alloy airframes.

In all these situations, a number of restraining factors must be considered:

Cost: In the majority of cases, new technologies decrease in cost as they become more established and are produced on a greater scale. From the outset, healable materials could also offset a higher initial outlay by increasing longevity and lowering (or even serving to eliminate) routine maintenance costs, which are especially high for safety critical situations such as transport.

Performance compared to conventional materials: Introducing healabilty into a functional component must not compromise the strength or increase the density of the component. A healing polymer that is double the density of its conventional (non-healing) predecessor will not be suitable for many applications, for example transport, where the energy required to move the heavy healing system will be seen to outweigh any potential safety advantage. This is particularly important when considering the mode of healing at the molecular level. Conceptually, introducing latent, unpolymerised monomers and functionalities into a material that can form new bonds in order to heal a fracture is a pleasing potential solution, but one must consider how much stronger the material would have been if the latent functionalities were fully polymerised in the original formulation.[7]

1.3   Categorisation of Healable Materials

As discussed, the simplest criterion of a healable material is that it is able to regain its strength after a damage event. This can be visualised by plotting strength of the polymer as a function of time (Figure 1.1). In this simplified example, the sample remains undamaged between time points A and B, indicated by the constant strength of the system. At time point B, the material is damaged, resulting in an instantaneous loss of strength (point C on the plot). Subsequently, the sample heals, and the strength of the material increases to match that of the pristine, undamaged sample (time point D). The rate of healing is indicated by the gradient of the slope between time points C and D. After healing, the strength remains constant for the remainder of the time.

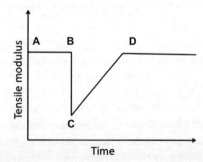

Figure 1.1   A schematic plot of tensile modulus as a function of time for a healable material undergoing a single break and heal cycle.

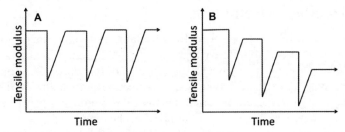

Figure 1.2 Schematic plots of tensile modulus as a function of time for healable materials undergoing multiple break/heal cycles: plot A) material that can completely recover the strength of the pristine material; plot B) material that regains only a fraction of the strength lost during the damage event.

A truly ideal healable material may have additional properties. For example, it may be able to heal on multiple occasions (Figure 1.2). These break/heal cycles may have several effects on the material. Figure 1.2A shows the strength *versus* time plot for an ideal sample that can regain its strength completely over three break/heal cycles. Figure 1.2B shows the situation encountered more frequently, whereby the sample only partially recovers its strength after each damage event. Thus, ultimately, the performance of sample will degrade, but, depending on the timescales involved, this may be adequate to fulfill the typical lifecycle of the product.

This simple treatment immediately raises several pertinent issues concerning healable material characterisation, some of which are highlighted:

   i) How strong is the pristine material?
  ii) How significantly can the material withstand damage yet still perform adequately with respect to its use?
 iii) How quickly can it regain its strength?
  iv) Under what conditions does the material heal?
   v) How many break/heal cycles can the material undergo? and
  vi) How easily can the site of damage in the polymer-based assembly be accessed?

The targets values for many of these questions will be application dependent. For example, the strength and number of healing cycles that a mobile phone case may be expected to pass through in a typical lifetime will be very different from that of a polymer component embedded deep within the wing of an aeroplane. However, a key universal consideration is what parameters should be measured to quantify healing in a polymeric system, and how they should be reported. The following two sections provide an overview of these parameters from a synthetic chemistry perspective, to provide an insight into some of the most common techniques and measurements reported for this rapidly evolving field of research.

1.4   Healing – Definitions

Although the initial design criterion for a healable material is the production of a component that can completely regain its physical properties after sustaining damage, there are many practical situations where this may not be possible or indeed necessary. In these situations, it is necessary to quantify the loss in performance of the healed material. This is most frequently expressed as the 'healing efficiency' (η_{eff}) of the material. The healing efficiency is a dimensionless parameter; it is the ratio of a specific mechanical property of the material before and after healing and is expressed as a percentage. It is given by Equation (1).

$$\text{Healing Efficiency, } \eta_{\text{eff}} = \left[\frac{\text{Mechanical Value (healed)}}{\text{Mechanical Value (pristine)}} \right] \times 100 \qquad (1)$$

Thus materials with η_{eff} approaching 100% (note: it is possible to have η_{eff} >100%) have regained the strength of the pristine material, whereas a low value for η_{eff} indicates that the material has not regained significant strength. Healing efficiency is therefore an easy parameter to obtain and clearly indicates the success of a healing process, but the simplicity of this term can lead to misleading results, especially when only a single healing parameter is measured. In a typical break/heal experiment, a sample will be stretched to its breaking point and a variety of mechanical properties measured, for example: tensile modulus, elongation to break and modulus of toughness. After healing, the sample may, and indeed frequently does, exhibit a higher η_{eff} value for one of these parameters when compared the other two. Thus only by measuring and quoting η_{eff} for all of the appropriate parameters for a specific polymer can a true assessment of the nature of the healable material be made.

A further complication for a researcher new to this relatively young field is that there is not a standard definition of what actually constitutes a *damaged* material. In this respect, papers describe η_{eff} values that may have been calculated for the healing of a single microscopic crack in a bulk polymer, or, alternatively, after a material has been physically broken into multiple sections and the parts then separated and rejoined prior to obtaining the healing efficiency data. To this end, the term 'healing' serves to cover several repair scenarios and is evidently subject specific; readers of articles in this area must always bear this caveat in mind.

1.5   Measuring Healing

Testing procedures for healable polymers have been inspired by the well-established areas of materials chemistry and engineering and broadly fall into two categories:

  i) mechanical load testing where a monotonic force is applied;
  ii) rheological tests where an oscillating force is applied.

Both these experimental protocols have been used to assess the healing properties of new polymeric materials. The most frequently studied of these techniques are covered in the following sections and also in Chapter 2.

1.5.1 Cantilever Beam Tests to Determine Healing

Cantilever beam (CB) tests are suited to testing non-elastomeric (brittle) samples, generally where the glass transition temperature (T_g) of the material is significantly above the testing temperature, or the polymer is highly cross-linked.[17]

Mechanical load testing requires fabrication of test samples with typical maximum dimensions of the order of $100 \times 75 \times 7$ mm (typically requiring 50 g of material). The geometry of the sample depends on the precise experiment (see Figure 1.3), with the four most common forms being:

   i) tapered double cantilever beam (TDCB);
  ii) compact tension (CT);
 iii) single edge notch bend (SENB) and
  iv) single edge notch tension (SENT).

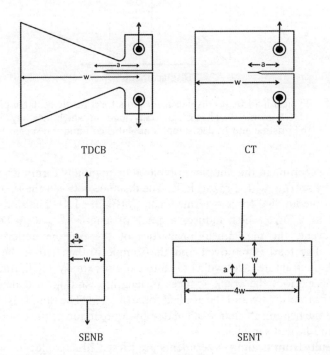

Figure 1.3 The four most common geometries for carrying out material property analysis on polymers. Key dimensions are indicated with double headed arrows: 'a' is the pre-crack length; 'w' is the most important geometric dimension. The position and direction of forces are shown with single headed arrows. Note: in the SENT experiment, the total force on the top and bottom face of the material must be equal.

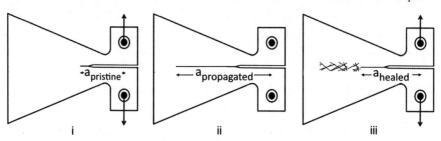

Figure 1.4   Typical break/heal experiment carried out on a sample with TDCB geometry: i) pristine sample with force being applied; ii) broken sample, crack has been extended and force removed to allow healing; iii) the sample has healed, the crack reduced in length and the force is reapplied to measure the strength of the healed material.

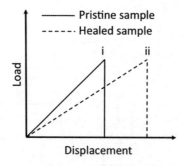

Figure 1.5   Typical load *versus* displacement plots for a brittle healable plastic in the TDCB geometry. The critical loads (loads at which they have broken) for the pristine and healed samples are labelled i and ii respectively.

In each experiment, the sample is prepared by manually forming a pre-crack, typically by scoring with a razorblade. The distance between the tip of the pre-crack and the applied force is termed $a_{pristine}$ (Figure 1.4). The sample is then subjected to a force which induces a crack of length $a_{propagated}$. During this fracture event, the mechanical properties of the pristine material can be obtained. The load is removed and the sample is then left to heal. Whilst unloaded, the fracture faces will come into close proximity, facilitating healing; during this period, the crack reduces in length, resulting in a new distance between the fracture tip and the applied force (a_{healed}). The sample is then tested again and the healing efficiency calculated by comparison of the response of the pristine and healed samples.

The results from healing experiments are presented as a plot demonstrating how applied load varies as a function of displacement, which is generally only a few millimetres. Representative load *versus* displacement plots for a break/heal experiment from samples in the TDCB geometry are shown in Figure 1.5.

It is evident from Figure 1.5 that the applied load is directly proportional to displacement, up to a critical value, for both the pristine and healed samples; beyond these values the samples break (as indicated by i and ii, Figure 1.5). The energy required to break the sample (defined as the *modulus of toughness*) is calculated by integration of the area under this displacement–load curve. In this typical example, where $a_{pristine} < a_{healed}$ (for example, see Figure 1.5), the healed sample exhibits a greater displacement before breaking. This results in an apparent increase the modulus of toughness for the healed sample and a corresponding η_{eff} of greater than 100%. However, this result is a consequence of the increase in the length of the pre-crack between the two tests, which results in an increase in compliance in the sample.

In the past decade, the TDCB geometry has become the method most commonly employed for investigating materials capable of exhibiting healable characteristics. This is primarily because the critical load (a typical healing parameter and indicated by points i and ii in Figure 1.5,) for the pristine and healed systems is independent of the difference in length between $a_{pristine}$ and a_{healed} (as shown in Figure 1.5). In each of the other three testing geometries (CT, SEND and SENT), $a_{pristine}$ and a_{healed} must be measured to a high precision and any discrepancies accounted for. If these time consuming and difficult measurements are not made and $a_{pristine}$ is approximated to be equal to a_{healed}, large errors (*i.e.* > 100%) are introduced into the results.

Load testing may also be carried out on elastic materials. In these cases, the sample is formed into a typical 'dogbone' like shape, requiring from 50 mg to multiple hundreds of grams of material depending upon the instrument used (see Figure 1.6). The increase in size of the sample at the termini serves to ensure that the sample breaks cleanly in the middle, preventing the sample tearing where it is clamped to the instrument. Furthermore, sample fracture at the clamped edges does not afford a representative indication of the strength of the sample, and thus sample preparation is key to this assessment format. The dogbone is elongated and the response of the material measured and plotted as

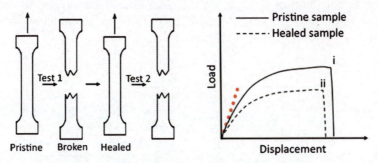

Figure 1.6 Left: Schematic of a break/heal test on a dogbone shaped elastomer. The pristine sample is stretched to breaking point and the response measured (Test 1). The broken sample is healed and re-tested (Test 2). Right: Typical displacement *versus* load plot for a pristine elastomer and healed sample. The ultimate tensile strengths of the two samples are labeled i and ii respectively. The Young's Modulus of the sample is measured by taking the tangent to the initial slope (dotted red line for the pristine sample). The Modulus of Toughness is the area under the curves.

stress (load) *versus* displacement (Figure 1.6). Extension values for typical elastomeric materials can be of the order of several times of the length of the pristine sample. The stress *versus* displacement plots reveal three key pieces of information about the sample which have been used as a measurement of healing efficiency:

i) the *ultimate tensile strength*, where the material breaks and the load response drops to zero;
ii) the *stiffness*, or *Young's Modulus* of the sample is measured by taking the tangent to the initial slope (see the dotted red line for the pristine sample in Figure 1.6);
iii) the energy required to break the sample, which is area under each curve (referred to as the *modulus of toughness*).

1.5.2   Impact Testing

Impact testing is a well-studied technique for assessing the strength of both metals and polymers. During the test, the sample is completely severed by a falling or swinging weight, from which the change in momentum of the system can be used to calculate the strength of the sample. Two of the most widely used variants of these experiments are the Izod and Charpy impact tests.[18,19] Sample sizes and geometries are of the order of $10 \times 10 \times 55$ mm and are defined by specific North American standards (for example ASTM D256 or A370). In both cases, a notched, clamped sample is struck and broken by the head of a swinging pendulum, and the loss of energy of the system is calculated by the difference in the height (angle) between the position that the pendulum began its motion and the angle it stops after breaking the sample. The two tests differ in that in the Izod test, the sample is clamped at one end of the long axis of the cuboid sample and the notch is on the same side as the impact. In contrast, samples for Charpy tests are clamped on both sides of the notch, which is on the opposite face to the impact (Figure 1.7).

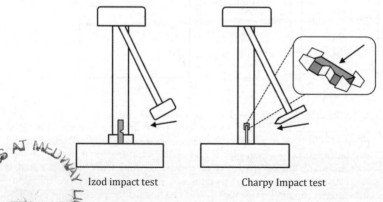

Izod impact test   Charpy Impact test

Figure 1.7   Schematics of the experimental set-up for Izod and Charpy impact testing.

The break/heal tests detailed in this section are conceptually easy to place into a real world setting. In each test, sample of bulk polymer is fractured or severed by applying a force, the sample is then left to heal and the material re-tested. Many polymers, however, including linear polymers and supra-molecular systems, are *viscoelastic* materials and as such are amenable to analysis by rheological techniques.

1.5.3 Rheological Healing Methods

Rheology is the study of the deformation and flow of matter and is under-pinned by an extensive mathematical framework, detailed in numerous excellent texts that are beyond the scope of this chapter.[20] The following sections will cover the important definitions, concepts and experiments that have been used widely to characterise and assess healable polymer systems.

1.5.4 Rheology – Background

All polymeric products are subject to forces of varying degrees throughout their lifecycle. It is therefore important to be able to model the strain (*i.e.* the displacement or change in shape) of these materials under a particular stress (*i.e.* the force applied to the polymer) – see Figure 1.8.

From the dawn of the modern polymer industry it was apparent that polymers in the bulk, melt and solution states did not display typical rela-tionships between force and deformation. For example, many solids are ideally elastic in nature and obey Hooke's law in that once the strain is removed, the materials recover their initial shape. Equation (2) demonstrated this for small strains (deformations). Conversely, many fluids (encompassing liquids and gases) follow Newtonian laws, whereby as the rate of strain increases, the stress (or drag) exerted by the fluid increases. This is shown by Equation (3).

$$\tau = G\gamma \tag{2}$$

$$\tau = \eta(d\gamma/dt) \tag{3}$$

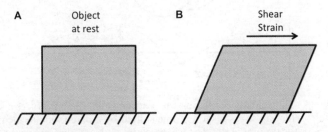

Figure 1.8 Schematic showing the strain response (displacement) of an object subjected to a shear strain.

Where τ = stress, force per unit area; G = elastic modulus, a constant of proportionality; γ = strain (change in length), thus $(d\gamma/dt)$ is rate of change in strain; and η = viscosity, a constant of proportionality.

These two laws are inapplicable to polymers, which frequently exhibit a response to stress that is intermediate between those predicted by these two simple equations; as such they are termed *viscoelastic* materials. The properties of viscoelastic materials are investigated by rheological experiments. During these tests, an oscillatory force with a frequency of ω is applied to one face of the material (solid line, Figure 1.9A), and the response to this force is measured at the second, fixed face of the material (dashed line, Figure 1.9A). The stress response lags behind the strain by a measurable time period, termed δ. Thus the stress and strain can be expressed as follows by Equations (4) and (5) respectively.

$$\gamma = \gamma_0 \sin \omega t \tag{4}$$

$$\tau = \tau_0 \sin(\omega t + \delta) \tag{5}$$

For analytical purposes, the stress response is decomposed into two waves, termed τ' and τ'' (Figure 1.9B), which sum to the same amplitude as τ [Equation (6)]. τ' is defined to be in phase with the strain input and therefore oscillates according to equation 4 (sin ωt), where as τ'' oscillates 90° out of phase with the strain (cos ωt) [Equation (6)].

$$\tau = \tau' + \tau'' = \tau'_0 \sin \omega t + \tau''_0 \cos \omega t \tag{6}$$

These two waves, τ' and τ'', allow for two new dynamic moduli to be defined: the elastic or storage modulus (G'), which is in phase with the shear strain, and the loss modulus (G''), which is out of phase with the shear strain input energy [Equations (7) and (8) respectively].

$$G' = \tau'_0/\gamma_0 \tag{7}$$

$$G'' = \gamma''_0/\gamma_0 \tag{8}$$

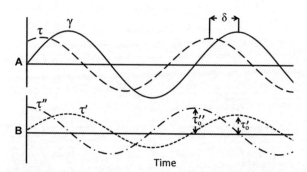

Figure 1.9 A) Sinusoidally oscillating shear strain (solid line) produces a sinusoidally oscillating stress response time shifted by a magnitude termed δ; B) Plots of the decomposed stress waves τ' and τ''. Note: these waves have the same frequency as γ, with one in phase (τ') and one out of phase (τ'') with the input strain.

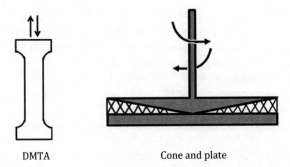

DMTA   Cone and plate

Figure 1.10 Example geometries for samples undergoing rheological healing tests. In the cone and plate rheometer, the polymer is placed in the hashed region.

Although these new moduli (G′ and G″) may appear to be mathematically abstract constructs, the storage modulus is a measure of the elastic energy retained in the system upon deformation, whereas the loss modulus accounts for the energy dissipated by the sample over each cycle. Thus by observing how these parameters vary as a function of either temperature or strain for samples before and after fracture, it is possible to build up a detailed picture of the healing behaviour of polymer samples.

Rheological healing tests using two geometries have been successfully conducted on materials with two geometries. The sample may be formed into a 'dogbone' shape which is subject to a linear extension and compression over a defined temperature range: dynamic mechanical thermal analysis (DMTA). Alternatively, the sample may be pressed between a flat base plate and cone shaped top plate which oscillates in a circular manner: cone and plate rheometry (Figure 1.10). Both measurements may be carried out on quantities varying from a few tens of milligrams to many hundreds of grams, depending on the equipment used.

In each test, the polymer sample is subjected to the oscillating strain until fracturing occurs. This is observed readily in the DMTA test as the severing of the dogbone sample. Samples tested in the cone and plate geometry are defined as broken when the recorded storage and loss modulus drops to essentially zero. The broken samples are then subjected to the designed healing stimulus (if required) and re-tested. Comparison of the storage and loss modulus of the material before and after healing allows the healing efficiency of the samples to be assessed. In the case of cone and plate rheometric tests, the broken sample may be left in the instrument for varying time periods prior to re-testing, as by the nature of the experiment, the 'fracture faces' are in contact.

1.6 Summary

This chapter has provided an overview of the key concepts required to study healable polymeric materials. Many of these techniques—for example, impact testing and rheometry—are far beyond the expertise of most synthetic chemists, who produce the materials to be analyzed. However, it is hoped that this brief introduction will help narrow the gap in understanding between chemists and materials scientists, further opening up the field of healable materials research to a new generation of scientists.

The following chapters in this monograph provide a reference text on the chemistry of healable polymers and cover, to date, the main chemical strategies used to afford such materials. Each of the following four chapters provides a critical assessment of the main developments in healable polymer systems:

Encapsulation-Based Self-Healing Polymers and Composites
This chapter provides detailed coverage of the development of the encapsulated monomer approach to healable polymer composites and also includes a summary of mechanistic aspects, with specific reference to key examples from the literature.

Reversible Covalent Bond Formation as a Strategy for Healable Polymer Networks
Reversible covalent bond formation processes such as Diels-Alder cyclo-additions have been utilized successfully to access healable polymer networks. This chapter will examine the range of reversible covalent bond formation processes that have used to this end.

Healable Supramolecular Polymeric Materials
Recent developments in healable polymer systems have employed key concepts from the field of supramolecular chemistry to afford exciting materials that can be repaired with high efficiency and reproducibly. Use of non-covalent inter-actions such as electrostatic forces of attraction, hydrogen bonding, metal-ligand binding and π-π aromatic stacking have been used to assemble polymeric networks built from relatively low molecular weight components – the hypothesis underpinning the healing characteristics of these materials is that disruption of weak non-covalent interactions facilitates repair prior to fracture of the covalent linkages in the oligomers.

Thermodynamics of Self-Healing Reactions and their Application in Polymeric Materials
This chapter will provide a detailed coverage of the thermodynamic considerations of healable materials. The role of the entropic and enthalpic contributions to the Gibbs free energy during self-healing events will be investigated, with a specific study of how the configurational changes of the polymers and chemical reactions within the material contribute to driving healing in bulk materials.

References

1. P. Sparke (ed.), *The Plastics Age: From Bakelite to Beanbags and Beyond*, Overlook Press, 1994.

2. For an essay concerning Staudinger's life and achievements, see: R. Mülhaupt, *Angew. Chem. Int. Ed.*, 2004, **43**, 1054.
3. For a contemporary review of his own work, see: W. H. Carothers, *Chem. Rev.*, 1931, **8**, 353.
4. Source: European Chemical Industry Council: http://www.cefic.org/ Facts-and-Figures/Chemicals-Industry-Profile. Retrieved June 2012.
5. S. van der Zwaag (ed.), *Self Healing Materials: an Alternative Approach to 20 Centuries of Materials Science*, 1st edn, Springer, Dordrecht, 2007.
6. S. K. Ghosh (ed.), *Self Healing Materials: Fundamentals, Design Strategies and Applications*, J. W. Wiley and sons Ltd., Weinheim, 2009.
7. D. Wagg, I. Bond, P.Weaver and M. Friswell (ed.), *Adaptive Structures: Engineering Applications*, J. W. Wiley and sons Ltd., Chichester, 2007.
8. R. P. Wool, *Soft Matter*, 2008, **4**, 400.
9. S. D. Bergman and F. Wudl, *J. Mater. Chem.*, 2008, **18**, 41.
10. Y. C. Yuan, T. Yin, M. Z. Rong and M. Q. Zhang, *eXPRESS Polym. Lett.*, 2008, **2**, 238.
11. S. Burattini, B. W. Greenland, D. Chappell, H. M. Colquhoun and W. Hayes, *Chem. Soc. Rev.*, 2010, **39**, 1973.
12. T. C. Mauldin and M. R. Kessler, *Int. Mater. Rev.*, 2010, **55**, 317.
13. E. B. Murphy and F. Wudl, *Prog. Polym. Sci.*, 2010, **35**, 223.
14. J. A. Syrett, R. C. Becer and D. M. Haddleton, *Polym. Chem.*, 2010, **1**, 978.
15. http://www.nissan-global.com/EN/NEWS/2005/_STORY/051202-01-e.html, last accessed 07/01/13.
16. www.bayercoatings.de, last accessed 07/01/13.
17. For an excellent review of CB testing with respect to healing materials, see: N. Brown, *J. Strain Anal.*, 2011, **46**, 167
18. S. H. Avner, *Introduction to Physical Metallurgy*, 2nd edn, McGraw-Hill, New York, 1974.
19. E. C. Rollason, *Metallurgy for Engineers*, 4th edn, Edward Arnold, London, 1973.
20. C. W. Macosko, *Rheology: Principles, Measurements and Applications*, Wiley-VCH Inc, New York, 1994.

CHAPTER 2

Encapsulation-Based Self-Healing Polymers and Composites

MICHAEL W. KELLER

Department of Mechanical Engineering, The University of Tulsa, 800 S. Tucker Drive, Tulsa, OK 74104
Email: mwkeller@utulsa.edu

2.1   Introduction

In the 1980s, researchers demonstrated the repair of polymeric materials *via* the addition of solvents or adhesives.[1] However, the approaches used lacked the ability to initiate and complete the repair automatically. Some form of outside intervention, either through the addition of solvent or through the application of heat or pressure, was required. Autonomy was introduced when White *et al.* demonstrated the first microcapsule-based self-healing material.[2] The introduction of microcapsules enabled the synthesis of a truly self-healing material that initiated and completed self-repair without any outside intervention. The capsules provided both stable storage and reliable release and were the transformative advance that made autonomic healing possible. Since that initial demonstration, the use of encapsulation as the basis for introducing self-healing functionality has expanded rapidly.

The specific details of any given encapsulation-based self-healing material may differ, but the basic mechanism for the incorporation and activation of the functionality remains the same. An encapsulated healing agent is embedded in a

RSC Polymer Chemistry Series No. 5
Healable Polymer Systems
Edited by Wayne Hayes and Barnaby W Greenland
© The Royal Society of Chemistry 2013
Published by the Royal Society of Chemistry, www.rsc.org

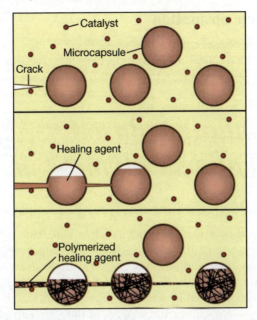

Figure 2.1   Schematic representation of a microcapsule-based self-healing material. Reprinted with permission from reference [2].

host matrix, along with either a catalyst or a cross-linker phase. Damage, typically a crack, then propagates through the material and ruptures the embedded capsules. The released healing agent is pulled into the damage by capillary action, where contact with a catalyst or a released cross-linker phase initiates polymerization. This initiator/activator phase has been both a solid that is dissolved by the released healing agent and a liquid that requires some degree of diffusive mixing. After contact with the catalyst/cross-linker, the healing agent polymerizes and bonds the crack faces together. The new adhesive bond is what restores the original mechanic properties of the host matrix. Schematically, this process of damage, response, and healing is shown in Figure 2.1.

Critical to the success of encapsulation-based self-healing materials is the interaction of the damage/fracture with the embedded capsules. One of the advantages of the microcapsule approach presented by White and co-workers is that cracks are attracted to the midplane of a microcapsule when the modulus of the capsule is less than that of the surrounding matrix.[2] An Eshlby-Mura equivalent inclusion model showed that the interaction of the stress field around an off-center crack and an inclusion would tend to drive the crack to the centerline of the microcapsule.[2] This prediction was confirmed by experiment.

In this chapter, the synthesis and characterization of microcapsule-based self-healing polymers is discussed. The microencapsulation processes that have been developed for the encapsulation of healing agents and the relevant healing chemistries are covered. Finally, a review of the testing and performance of self-healing polymers and composites is presented.

2.2 Microencapsulation

The encapsulated phase is the critical component of this class of self-healing materials. Capsules are the self-healing mechanism, serving both as storage and trigger of the healing response. Consequently, a significant volume of literature reports on the encapsulation of a wide variety of chemically active agents intended for use in self-healing materials. These encapsulation approaches draw on the related literature that is devoted to encapsulation of materials ranging from fragrances to pharmaceuticals. Several comprehensive reviews of encapsulation exist,[3,4] but a few important techniques that are frequently used in the context of self-healing materials will be briefly described.

2.2.1 *In situ* Poly(urea-formaldehyde) Encapsulation

Encapsulation in a poly(urea-formaldehyde) or melamine formaldehyde shell has proven to be one of the most versatile and widely used encapsulation approaches within self-healing. In this chapter, these systems will be referred to by the abbreviation, UF. The UF encapsulation process is emulsion-based and requires that the encapsulent be relatively immiscible in water. A detailed study of the encapsulation procedure and analysis of the microcapsule formation was given in a paper by Brown *et al.*[5] The basic procedure for this encapsulation is shown in Figure 2.2.

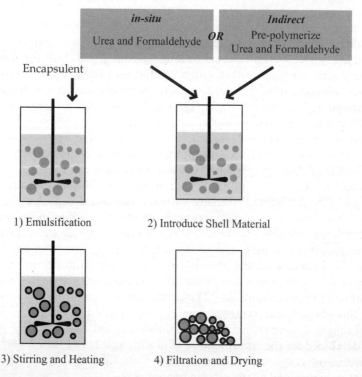

Figure 2.2   Schematic of the UF encapsulation procedure.

Microcapsules are produced by the polymerization and deposition of the UF polymer at the interface of the suspended emulsion droplet. Polymerization of the shell wall occurs in the aqueous phase until a critical molecular weight is achieved and the polymer phase-separates, depositing at the encapsulent–aqueous interface. The deposited polymer forms the shell wall of the microcapsule.

Two broad variations on this microencapsulation approach have been reported in the literature. The *in situ* technique uses direct addition of urea and formaldehyde to the aqueous phase of the encapsulation bath. An emulsion of the encapsulent is established and urea and formaldehyde are then added to the aqueous phase. Adjusting the pH to slightly acidic and heating the encapsulation bath initiates and completes the encapsulation. Capsules produced using this technique typically have thin shell walls, of the order of a few hundred nanometers, with a relatively thick porous layer of UF nanoparticles attached to the surface.[5] However, if the encapsulent to wall material ratio is high enough, almost the entire mass of shell wall precursors can be consumed to produce shell wall, resulting in microcapsules with surfaces that are relatively smooth.

An indirect technique can also be used to produce UF microcapsules. In this approach, a prepolymerization of urea, melamine, or phenol and formaldehyde is performed in a separate reaction vessel. The prepolymer is then added to the emulsified core material at the point where the urea and formaldehyde would normally be added in the *in situ* process. Encapsulation proceeds as it would in the *in situ* process. Capsules formed using the indirect procedure can have much thicker shell walls, up to several microns.[6]

Emulsions for both encapsulation approaches are typically produced using physical stirring. The mixer blades induce a strong shear flow that breaks the encapsulent into small droplets. Capsule diameter is then directly relatable to stirring speed. As an example, Figure 2.3 shows the diameter of dicyclopentadiene (DCPD)-filled microcapsules produced using the *in situ* technique.

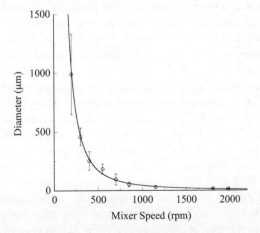

Figure 2.3 Size of DCPD-containing microcapsules as a function of stirrer speed. Reproduced with permission from reference [6].

While the mechanical agitation approach allows for the production of capsules as small as 20 to 30 microns in diameter, the ability to generate capsules of a single micron or less is generally beyond the scope of this technique. Several approaches to synthesizing sub-micron capsules have been presented in the literature. The first demonstration of sub-micron micro-capsules utilized a combined ultrasonic dispersion and stirring technique to generate an emulsion with nanometer-sized droplets.[7] Capsules were synthesized by irradiating a low volume capsule batch with ultrasound and then stirring to maintain the dispersion during standard *in situ* UF encapsulation. This process was capable of producing capsules of between one micron and several hundred nanometers in diameter.

Other researchers have used the Shirasu Porous Glass method to generate an emulsion with a narrow size distribution.[6] Emulsions are produced by forcing an immiscible liquid through the membrane into a host liquid. These systems have demonstrated the capability to produce almost monodisperse liquid and gas emulsions. In an encapsulation setting, the membrane is used to form the droplets in the emulsion for an indirect UF encapsulation. One of the novel features of the reported procedure was the construction of a rotating membrane that served as both the emulsion source and the mechanical agitator that maintained the emulsion. While this approach yielded capsules that were within the capability of mechanical stirring, about 40 microns, the distribution of diameters was much narrower when compared to mechanically emulsified encapsulation procedures.

In addition to direct formation of microcapsules filled with a core of healing agent, the UF encapsulation procedure has also been successful in manu-facturing capsules through a diffusion-based process. Encapsulation *via* this method is achieved by first producing either hollow or solvent-filled capsules. As mentioned previously, the indirect UF procedure produces capsules with relatively thick shell walls. These capsules are therefore far more stable to buckling collapse when compared to the capsules produced by the *in situ* technique and can be used to encapsulate air or gas bubbles entrained in the encapsulation bath. These bubbles can be produced by including a blowing agent in the encapsulation bath or by aggressively stirring an encapsulation bath to entrain air. The hollow capsules can then be filled using a diffusion approach. Capsules are dispersed in the liquid of interest, which diffuses inward through the capsule shell wall, eventually filling the capsule as osmotic pressure equalizes. This approach has proven useful for encapsulating highly reactive or water sensitive materials such as boron trifluoride diethyl etherate.[8]

One continuing concern for the microcapsule-based approach is the stability of the capsules. The diffusive mobility of the core material across a thin polymer shell is a benefit for manufacturing capsules with otherwise un-encapsulatable core materials, but also presents a problem when storing the capsules in a dry form. The stability of capsules for a small range of core materials was reported in a paper by Nesterova, Dam-Johansen, and Kiil.[9] In this paper, both the urea-formaldehyde and melamine-formaldehyde encap-sulations were performed with an epoxy resin as the core material. Capsules were found to be unstable after long term storage in air (approximately

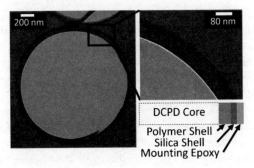

Figure 2.4 TEM image of a microtomed, silica-protected microcapsule containing DCPD.
Reprinted with permission from reference [10].

7 weeks). To address these stability issues, researchers have developed two-step encapsulation procedures that synthesize capsules with two shell walls. One of the first reports of this was the silica coating of an existing UF-walled, sub-micron capsule containing dicyclopentadiene.[10] A cross-section of the capsule is shown in Figure 2.4.

Capsules were coated using a fluoride-catalyzed condensation reaction using tetra orthosilicate as the starting material. A benefit, aside from the added stability, was improved resistance to agglomeration, which made smaller capsules easier to handle.

2.2.2 Interfacial Polymerization

The second most popular encapsulation procedure for self-healing materials is the interfacial polymerization approach. This is a fundamentally different process to the *in situ* UF or melamine-formaldehyde process. Instead of a polymerization reaction that proceeds mostly within the aqueous phase of the encapsulation bath, the polymerization reaction in the interfacial encapsulation procedure occurs at the encapsulent–water interface, building the shell wall by forming a film that thickens with time.

Interfacial polymerizations are typically step-growth reactions where one monomer is hydrophilic and the second monomer is hydrophobic. An inter-facial polymerization is classically demonstrated in the reaction of a diamine in aqueous solution and a sebacoyl chloride in non-polar solvent. This arrangement is referred to as the "nylon rope trick," as a thin cord of nylon polymer can be drawn up from the film that forms at the interface of the aqueous and organic phases. In encapsulation, this is exploited by dissolving a hydrophobic prepolymer or monomer in the intended encapsulent. A co-solvent is sometimes required. This solution is then emulsified as with the UF system. An aqueous monomer is then added to the encapsulation bath and the polymerization proceeds at the interface between the encapsulate and the water, building the polymer around the droplet and forming the shell wall.

Interfacial encapsulations based on polyurethane chemistry have been the most common for self-healing materials. In a typical encapsulation, a prepolymer derived from an aromatic diisocyanate, typically toluene diisocyanate (TDI) or methyldiphenyl diisocyanate (MDI), is dissolved in the encapsulent phase. A co-solvent is frequently required to ensure that the prepolymer and encapsulent are miscible. The resulting solution is then emulsified in an aqueous bath that contains an emulsion stabilizer. Gum Arabic is commonly used, but other stabilizers have also been successful.[11,12] After the emulsion has stabilized, a cross-linker or chain extender is added to the aqueous phase. In the case of polyurethanes, diols such as ethylene diol and butanediol are typically used. Polymerization occurs at the interface between the encapsulent and the aqueous phase forming the shell wall (Figure 2.5). The microcapsules can have relatively thick shell walls (of the order of a few microns). However, unlike with the UF system, shell wall thickness can be controlled with reasonable fidelity by varying the reaction time or mass of the available shell wall precursors.

This approach has been used for a variety of healing agents, ranging from an organotin catalyst[11] to amine curing agents.[13] One of the more interesting approaches was to utilize the difference in reaction rate between aromatic diisocyanates and aliphatic diisocyanates to encapsulate a reactive isocyanate core. In the first demonstration of this encapsulation procedure, a toluene diisocyanate (TDI)-based prepolymer was used to encapsulate isophorone diisocyanate. The 3 orders of magnitude difference in reaction rates between the aliphatic isophorone diisocyanate (IPDI) and the aromatic TDI led to preferential TDI consumption and produced microcapsules containing a reactive isophoronecore.[14] Building on this approach, Yang and co-workers utilized a similar approach with an methyldiphenyl diisocyanate-based pre-polymer to encapsulate a hexamethylene diisocyanate core material.[15]

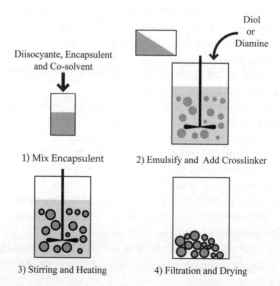

Figure 2.5 Schematic of the polyurethane interfacial encapsulation procedure.

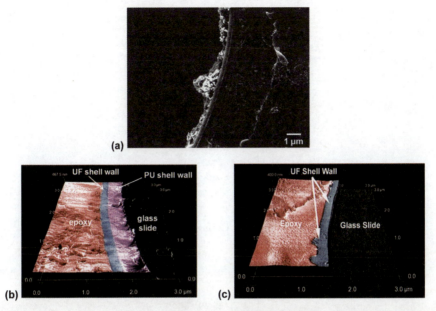

Figure 2.6 (a) An SEM image showing the double shell wall microstructure of the capsule. Images (b) and (c) are AFM images of a double wall microcapsule, image (b), and a single walled microcapsule, image (c). Reprinted with permission from reference [16].

An interesting combination of the two microencapsulation procedures described was used to produce a highly stable, double-walled microcapsule formed from an inner polyurethane (PU) shell microcapsule with an outer UF capsule in one step, unlike the two step encapsulations described above.[16] SEM and AFM phase images of the double capsule wall is shown in Figure 2.6. The single step production was accomplished by simply running both the *in situ* UF and interfacial polymerizations at the same time. Kinetic differences in the two polymerizations lead to the formation of a PU inner shell and a UF outer shell.

These double-wall microcapsules demonstrated exceptional stability, showing retention of core material after 1 year of storage, with little or no degradation. Furthermore, these capsules also demonstrated significantly better stability when exposed to high temperatures, as might be experienced during a composite cure cycle.

2.2.3 Hollow Fiber Encapsulation

An alternative to the broadly chemical microencapsulation schemes described previously is the use of hollow fibers as the container for a given healing agent. These fibers can be as simple as a heat-sealed glass pipette,[17] but the most successful systems use fibers produced from an advanced fiber-pulling system.[18] This approach takes a fiber preform in the desired fiber geometry and draws out the fiber that retains the initial shape of the preform, but at much reduced diameter. This is shown schematically in Figure 2.7.

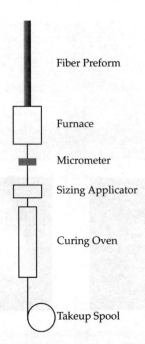

Figure 2.7 Schematic of the fiber pulling process. The fiber preform is loaded into a
             furnace that heats the material and allows a filament to be pulled to a
             smaller diameter.

Almost any fiber shape can be generated using this approach, but hollow
fibers, such as those shown in Figure 2.8, are employed in the same manner as
microcapsules. Fibers are simply incorporated into a composite assembly where
they serve as the repositories for healing agents. Filling of the fibers is
completed through a vacuum-assisted infiltration process that pulls the desired
core material into the embedded hollow fibers.[18] Currently, these fibers are not
prefilled prior to the fabrication of a composite material.

   In addition to glass fibers, multi-component polymeric fibers have also been
described in the literature.[19] To manufacture these multi-component fibers, an
emulsion of the target encapsulent and the fiber material was made and then
charged into a fiber spinning apparatus. The reported system utilized dichloro-
benzene (DCB) as the encapsulent and the fibers used were alginate. The fiber
was then drawn out and the emulsified droplets became encapsulated within the
fiber, as shown in Figure 2.9.

   Finally, an electrospinning-based encapsulation scheme has also been
demonstrated for the encapsulation of potential healing agents. Electrospinning is a modified fiber spinning process that combines standard polymer
jet spinning with a high voltage electrical field. The electrostatic forces
generated by the applied electric field creates a very narrow jet that projects
from the polymer solution, enabling the formation of thin polymer fiber. Final
fiber diameters can be of the order of a few hundred nanometers. Encapsulation
by electrospinning uses a coaxial approach where a set of nested spinning
needles is used to create a coaxial polymer jet with the shell material on the

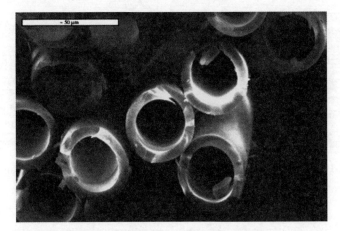

Figure 2.8 Hollow glass fibers produced using the fiber drawing process.
Photo courtesy of Professor Ian Bond of the University of Bristol.

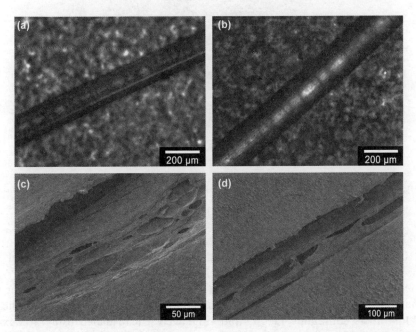

Figure 2.9 (a) and (b) Microscope images of a fiber containing compartmentalized;
(c) and (d) a microtome SEM showing the hollows in a mounted and
extracted fiber.
Reprinted with permission from reference [19].

exterior and the target encapsulent in the inner stream. In the case of liquid
inner cores, a beaded morphology is typically produced. Indeed, Park and
Braun[20] have successfully utilized a coaxial electrospinning process to produce
fibers with a beaded morphology, containing a poly(dimethylsiloxane) healing
agent, as shown in Figure 2.10.

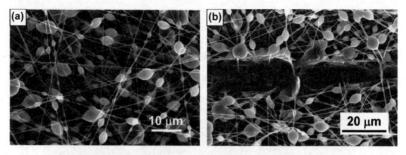

Figure 2.10   Beaded fibers containing a poly(dimethyl siloxane) healing agent. Fibers
were formed from a poly(vinylpyrrolidine) polymer shell.
Reprinted with permission from reference [20].

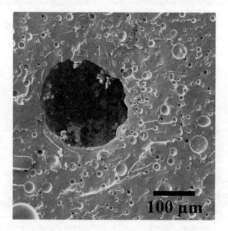

Figure 2.11   Fracture surface image showing phase separated domains and a
fractured microcapsule.
Reprinted with permission from reference [11].

In addition to siloxane, DCPD and IPDI have also been encapsulated by
electrospinning,[11] as shown in Figure 2.11.

Sinha-Ray *et al.* utilized a polyacrylonitrile shell to encapsulate the target
materials, but proceeded in a slightly different manner to previous researchers.
While fibers containing DCPD and IPDI were encapsulated using a coaxial
approach, as described previously, fibers were also fabricated using an emulsion
technique and a solution blowing process.[21]

In emulsion electrospinning, an emulsion of core and shell material is
prepared prior to electrospinning. This results in randomly distributed core
material housed within the electrospun fiber, similar to the microstructure
shown in Figure 2.10. In solution blowing, high-speed air jet is directed
coaxially around a spinning needle and induces a bending instability similar to
that of electrospinning. A fiber is synthesized without the requirement of
electrical conductivity to successfully electrospin. None of the fibers produced

in this study were utilized to produce a self-healing material, but the desired reactivity was confirmed.

2.2.4 Microcapsule Characterization

Microcapsule characterization encompasses a broad range of tasks, from simple confirmation of capsule formation to quantitative analysis of mechanical performance. The most basic characterization is the 'crush test'. Dried capsules are placed between two microscope slides and crushed to verify release. These tests, coupled with optical microscopy, are the primary methods by which capsule formation is ascertained. More detailed testing employs a range of techniques such as thermogravimetrtic analysis (TGA), differential scanning calorimetry (DSC), elemental analysis, electron microscopy, and mechanical testing, which will be described sequentially.

TGA is typically used to determine what percentage of a given capsule batch is core material and as a qualitative technique for ascertaining capsule quality. To perform this test, capsules (typically *ca.* 5 mg) are placed in a small crucible on a highly sensitive microbalance contained within a temperature-controlled furnace. Depending on the oxidative stability of the capsule and core, this test can be conducted in air or inert atmosphere; inert atmosphere is typical. As the capsules are heated, changes in mass are recorded and can be compared to control runs using pure core and shell materials. Alternatively, the TGA furnace can be coupled to a mass spectrometer to attempt quantitative analysis of the boil or burn off products. The shape of the TGA mass loss curve is frequently used to estimate capsule quality. Ideal capsules exhibit minimal or no mass loss prior to the boiling point of the core and then a large drop as the core material vaporizes and bursts the capsules. Poor capsules are associated with TGA traces that begin to lose mass continuously as soon as the heating cycle begins. A comparison of good and poor capsule TGA traces is shown schematically in Figure 2.12.

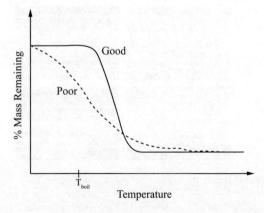

Figure 2.12 Schematic representation of a "good" (solid line) and a "poor" (dashed line) TGA trace of microcapsules.

DSC has been used by several groups to confirm reactivity of encapsulated materials.[22–24] This experiment is performed by placing a small amount of crushed capsules (typically *ca.* 5 mg) in a metallic crucible. The sample crucible and an empty reference crucible are placed in a furnace and analyzed using precise thermocouples. As the temperature is increased, any discrepancy between the temperatures of the two thermocouples is directly related to the heat generation caused by chemical or phase changes occurring within the sample crucible. Conceptually, this technique should be very powerful in determining the chemical activity of the encapsulated materials; however, phase

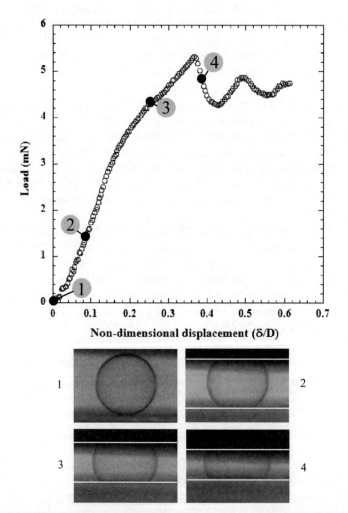

Figure 2.13   Load-displacement plot of the compression of a single microcapsule containing DCPD. Inset images are taken at the points indicated on the displacement curve, where 1 is pre-compression, 2 is during the elastic region, 3 is post 'yield', and 4 is post burst.
Reprinted with permission from reference [25].

transitions within the capsule shell, as well as side reactions, can mask or obfuscate what should otherwise be clear reaction peaks.

Mechanical testing of microcapsules has also been performed to investigate capsule stability and strength. Keller and Sottos adapted a microcompression experiment to investigate capsule burst strength and to extract a modulus of the shell wall for capsules produced by *in-situ* UF microencapsulation.[25] This experiment is performed by compressing a single microcapsule between rigid platens until failure. An example load-displacement plot of the compression of a UF capsule is shown in Figure 2.13. Burst strength is determined directly from the maximum load experienced during the test, and the modulus is extracted by comparison with one of several available models for capsule compression. This approach has been used to investigate the mechanical properties of micro-capsules formed by the UF process,[25] interfacial polymerization,[14] and by some of the more advanced processes described previously.[16]

Microscopy plays a key role in the characterization of microcapsules. Optical microscopy is typically utilized for capsule size distribution analysis, as well as confirmation of capsule formation. Most microcapsule shell walls are made of amorphous polymers and are therefore somewhat transparent, as can be seen in Figure 2.13. This transparency enables the use of fluorescent dyes to investigate capsule release behavior. However, most microcapsules are relatively large and the limited depth of focus of optical microscopes requires that scanning electron microscopy be used for more complete characterization of capsule morphology and shell wall thickness. Transmission electron microscopy is required for the analysis of the smallest microcapsules (*i.e.* <10 μm).

2.3 Chemistry of Encapsulated Self-Healing Materials

The main body of development on encapsulation-based self-healing materials has focused on the expansion of the range of available healing chemistries. Suitable healing chemistries have the following properties:

1. amenable to encapsulation;
2. stable during extended periods of inactivity;
3. relatively fast reaction rates, *i.e.* hours, not days;
4. near room temperature reactions;
5. one liquid constituent and
6. reaction is insensitive to air and moisture.

These six characteristics are general requirements for the selection of an appropriate healing chemistry. Note that the final characteristic is not so much a requirement as a guideline, as several approaches have used environmental moisture as the trigger for healing. The requirement of encapsulability is typically the most stringent and has prevented several seemingly ideal healing chemis-tries—such as the use of very reactive cyanoacrylates—from being adopted.

While the details of the performance testing and characterization of self-healing materials will be discussed later in this chapter, the concept of healing efficiency to help quantify the relative benefits and drawbacks of individual healing chemistries also merits coverage (see Chapter 1). The use of healing efficiency to quantitatively describe the effect of self-healing was first proposed by Wool and O'Conner.[1] Healing efficiency, η, is defined as the recovery of some defined material property; see Equation (1).

$$\eta = (\Gamma_{healed}/\Gamma_{initial}) \times 100\% \tag{1}$$

The particular material property can be fracture toughness, modulus, ultimate strength, or any other measurable material parameter. The techniques that have been developed to investigate healing efficiency will be discussed in detail in the next section and have also been outlined in Chapter 1.

2.3.1 Ring-Opening Metathesis Polymerization

Ring-Opening Metathesis Polymerization (ROMP) was the first healing chemistry used in a fully autonomic self-healing polymer. This system is composed of a cross-linkable monomer, DCPD, and a solid-phase polymerization catalyst (1st generation Grubbs' catalyst).[2] ROMP of DCPD proceeds as shown in Scheme 2.1. ROMP polymerization proceeds by attack of the Grubbs' catalyst on the *exo* face of the monomer and opens the strained double bond on the norbornene-derived double bond. This reaction forms the initial, linear polymer. The cyclopentadiene-derived double bond will then be consumed to produce addition polymer and to crosslink with other, growing chains.

This healing chemistry has several advantages for self-healing materials. The monomer, DCPD, is readily encapsulatable with the *in situ* UF process described previously and is stable to auto-polymerization and degradation over long periods. The healing agent is also a low viscosity liquid, which enables efficient crack plane coverage *via* capillary forces. From a cost standpoint, DCPD is also relatively inexpensive; fortuitously, it is a by-product of the steam cracking process. Grubbs' catalyst is relatively expensive, but only small quantities are required to successfully initiate and sustain ROMP.

The kinetics of the ROMP polymerization using 1st generation Grubbs' catalyst were investigated from a mechanistic approach by Rule and Moore.[26] In this study, the relative reactivity of the *endo* and *exo* isomers of DCPD, shown in Figure 2.14, were investigated.

In this study, the *exo* isomer was shown to polymerize faster than the *endo* form. Mechanistically, this reaction rate difference was attributed to steric

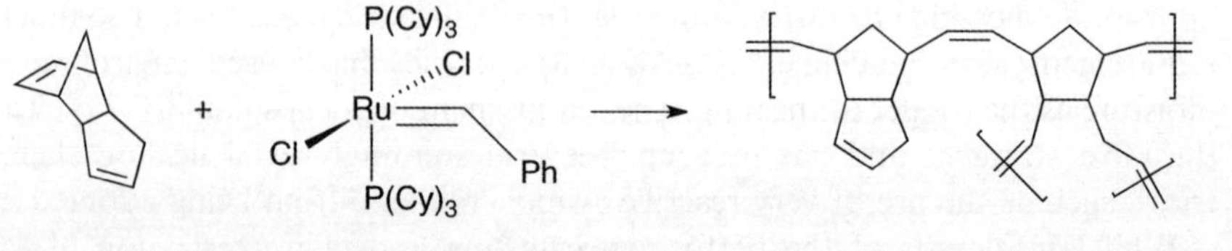

Scheme 2.1   ROMP of the cross-linkable monomer DCPD by Grubbs' catalyst.

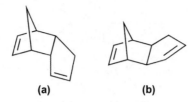

Figure 2.14 Isomers of dicyclopentadiene, (a) *endo* and (b) *exo*.

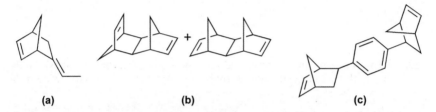

Figure 2.15 Healing agents studied as potential self-healing chemistries. Ethylene norbornene, (a), was blended with the cross-linkers (b) and (c).

hindrance of an approaching monomer by the ruthenium center. The kinetics of the polymerization reaction was studied from a cure kinetic approach by Kessler and White.[27] This study investigated the impact of catalyst concentration on cure rates and found, as expected, that greater concentrations of catalyst increased the cure rate and the ultimate degree of cure.

In addition to the studies of *endo* and *exo*-DCPD, other groups have explored new ROMP-active monomer blends as candidates for healing agents. Liu and co-workers investigated the use of ethylene norbornene (ENB) blended with two norbornene-derived cross-linkers that are shown in Figure 2.15.[28]

These monomer blends were evaluated using a novel rheokinetic technique, which was based on coating the surfaces of a parallel plate fixture with epoxy containing Grubbs' catalyst. Sanding the coating exposed the active catalyst and the target healing agents were then applied, and the curing polymer film was analyzed by rheology. This experimental set-up mimicked the conditions on the crack plane and allowed for a simulation of the real-world conditions of polymerization for a healing agent. Using this experimental approach, the addition of cross-linkers was found to significantly improve the cure time of the healing agents, but attempts were not then made to use these new healing agents in self-healing materials.

Several derivatives of Grubbs' catalyst have been developed since the introduction of the first generation. The ROMP activity of the 2nd generation Grubbs' catalyst and Hoveyda-Grubbs catalyst (shown in Figure 2.16) was investigated by Wilson and co-workers using DCPD and two other ROMP-active monomers, 5-ethylidene-2-norbornene and 5-norbornene-2-carboxylic acid (NCA). These catalysts exhibit improved thermal and environmental stabilities, key parameters for the production of long-lived healing materials.

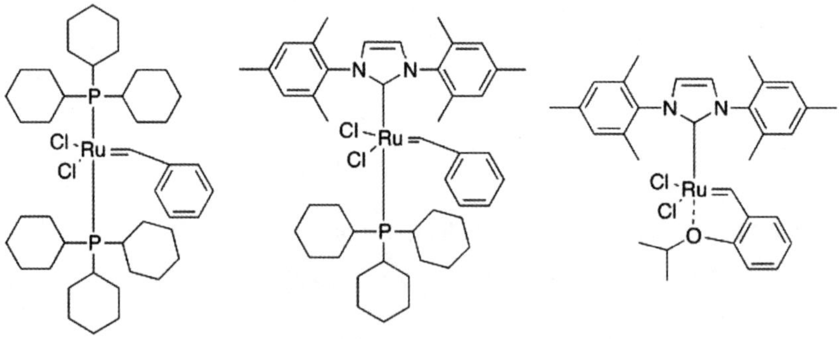

Figure 2.16   Variants of Grubbs' catalysts: A – 1st generation Grubbs' catalyst; B – 2nd generation Grubbs catalyst; and C – Hoveyda-Grubbs' catalyst.

In solution-based kinetic studies, the next generation catalysts performed significantly better. However, in mechanical testing of fully *in situ* healing, the next generation systems did not perform as well as the first generation catalyst. The primary reason for this discrepancy was found to be the competition between the dissolution of the catalyst and the polymerization of the surrounding monomer. The higher reaction rates for the second generation catalysts led to the formation of a cross-linked shell around undissolved catalyst particles. This shell prevented the catalyst from dissolving in the surrounding monomer and significantly limited the catalyst available for polymerization. As observed by Kessler and White, catalyst concentration was a critical parameter for determining rate and ultimate cure state of the DCPD/Grubbs healing system.[27]

Based on these results, Jones and co-workers used a model system to study catalyst dissolution and polymerization kinetics.[29] First generation Grubbs' catalyst with several physical morphologies was synthesized using a solvent exchange, a freeze-drying, and a precipitation approach. As shown in Figure 2.17D, catalyst particles as received from the supplier (in this case Sigma-Aldrich) are large cubical crystals (note the shift in scale bar in Figure 2.17D). These larger crystals do provide some benefit in that the larger particles allow for the formation of *in situ* encapsulation with a thin shell of deactivated catalyst surrounding and protecting a 'core' of active catalyst. Each of the different physical morphologies afforded different dissolution rates. The greatest dissolution rate observed following the freeze-dried morphology, Figure 2.17C, which produced of thin, porous wafers. Unfortunately, the large surface area to volume ratio for this catalyst form was more easily deactivated by curing agent contact.[29]

In addition to catalyst processing, Maudlin and Kessler studied the impact of the healing agent on the kinetics of catalyst dissolution in self-healing systems. Using a Hansen solubility parameter approach to investigate a series of

Figure 2.17 Comparison of catalyst morphologies based on various processing techniques: A – vacuum evaporation from CH_2Cl_2; B – precipitation from CH_2Cl_2/acetone; C – freeze-drying from benzene; and D – as received from Sigma-Aldrich. Note the scale bar difference in image D. Reprinted with permission from reference [29].

norbornene derivatives, these researchers were able to determine specific pendent functionalities that enhanced the dissolution of Grubbs' catalyst. However, the Hansen parameters were found to be non-predictive when comparing monomer blends with significantly different viscosities.[30]

In order to reduce the degree of deactivation of the catalyst by the amine curing agent in epoxy, Rule and co-workers developed a wax encapsulation procedure to protect the catalyst.[31] Ground catalyst was suspended in a molten paraffin wax that was then added to a heated, stirred water bath containing a surfactant. After the wax was emulsified, a cold water charge was added to rapidly lower the bath temperature and solidify the molten wax into catalyst-laden particles. These particles were then incorporated into a self-healing composite in the same manner as microcapsules or catalyst particles. The catalyst was activated by the healing agent dissolving the surrounding paraffin encapsulation and then dissolving the now exposed catalyst particle. Wax encapsulation protected the catalyst from the amine curing agent and allowed for an order of magnitude reduction in catalyst loading, from 2.5 wt% to 0.25 wt%, with little impact on healing efficiency. Introducing paraffin wax into the curing healing agent did change the mechanical behavior of the healed specimen, as wax is a plasticizer for poly(DCPD). The plasticization introduced non-linear fracture behavior, which will be discussed in greater detail.

Grubbs' catalyst-based ROMP systems are extremely flexible and effective in self-healing systems, but suffer from a significant drawback, the cost of the catalyst. Currently, the cost of Grubbs' catalyst is roughly twice the cost of an equivalent quantity of gold. While the quantity of catalyst required is relatively low, the cost can be significant for low-value, small margin composite and polymer consumer goods. In an effort to develop a cost-effective, ROMP-based, self-healing system, Kamphaus and co-workers investigated the use of a tungsten-based catalyst system as a direct replacement for the more costly Grubbs system.[32] The system was based on tungsten hexachloride with a phenylacetylene co-activator and required the use of the *exo* isomer of DCPD to achieve useful polymerization rates. Unfortunately, as a result of a combination of chemical incompatibility with epoxy resin and a tendency to clump in the matrix, healing efficiencies were considerably lower when compared to the original DCPD/Grubbs chemistry.

2.3.2   Siloxane-Based Healing Systems

Siloxane-based healing chemistries are divided into two classes of poly-merization systems. A polycondensation polymerization utilizes an alkoxy end-capped siloxane prepolymer and a hydroxyl end-capped cross-linker. This condensation reaction is initiated and mediated by one of several catalysts. Cho and co-workers explored this system, reported in several papers, using the polycondensation approach shown in Scheme 2.2.[11,33]

In this research, the catalyst was an organotin compound (dilauryldibutyl tin), which was encapsulated using an interfacial polyurethane procedure. The resin and cross-linking material were either encapsulated together within a single UF microcapsule, or mixed directly into the matrix material to form a phase-separated material, as described previously. Healing efficiencies for this system, when used to repair an epoxy matrix, were low, likely the result of the significant mismatch between the material properties of the host matrix and healing agent. Notably, however, this healing chemistry has demon-strated the capability of self-healing when immersed in water.[33]

Hydrosilylation was utilized by Keller and co-workers for the healing of an elastomeric poly(dimethylsiloxane). The hydrosilylation-based healing chemistries function by the cross-linking of a vinyl-functionalized prepolymer with a hydrosilane functionalized cross-linking agent. A platinum-based catalyst, typically Kardstedt's catalyst, is present in small quantities in one of the two liquid components of the healing system.

Scheme 2.3 shows the cross-linking reaction for this chemistry, which has two unique characteristics: i) healing was demonstrated in a chemistry that required the mixing of two liquid components on the crack plane and ii) the healing chemistry also produced the same polymer as the matrix polymer. The matching of the healing agent-derived polymer and the matrix material led to extremely high healing efficiencies.[34] Another feature of this healing system is that the released healing agent was able to take advantage of unreacted monomer, or latent functionality, present in the matrix to accomplish healing. Several healing systems have since attempted to utilize this same approach to improve healing efficiencies.[35]

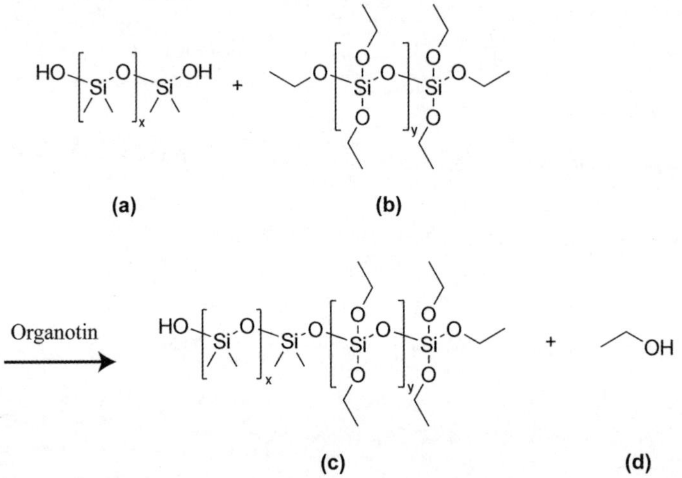

(a)　　　　　　　　　(b)

(c)　　　　　　　　　(d)

Scheme 2.2 Polycondensation of hydroxyl terminated siloxane 'a' and alkoxy end-capped siloxane 'b' used to produce a polymer system capable of healing whilst submerged in water.

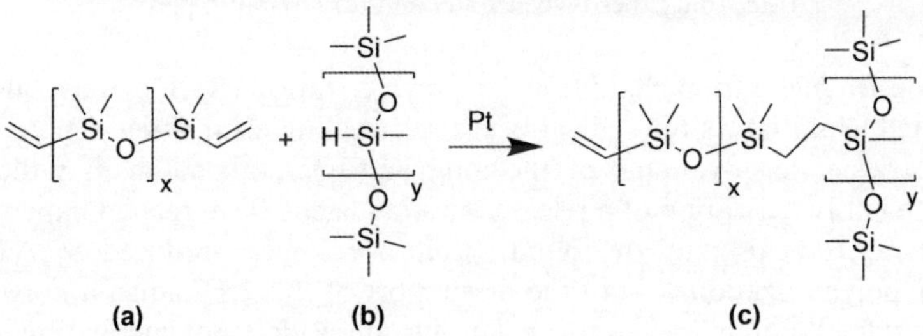

(a)　　　　　　　(b)　　　　　　　(c)

Scheme 2.3 Platinum-catalyzed hydrosilation chemistry used to produce a healable elastomeric siloxane based polymer.

2.3.3 Radical-Initiated Healing Chemistries

Wilson and co-workers were the first to investigate the potential of acrylate systems as healing chemistries by studying the kinetics of the various peroxide-based polymerization initiators.[36] In this initial study, five radical polymerization initiators were investigated as shown in Figure 2.18.

Of the peroxide initiators that were studied, two were amenable to encapsulation within a UF shell, benzoyl peroxide (BPO) and lauroyl peroxide (LPO). Microencapsulation was performed by dissolving the initiators in a solvent—phenyl acetate for BPO and hexyl acetate for LPO—and then encapsulated as described previously. Healing was investigated by control experiments, but an *in situ* system was not tested.

The active chain ends of polymers formed by living radical polymerizations, such as atom transfer radical polymerization (ATRP) and reversible

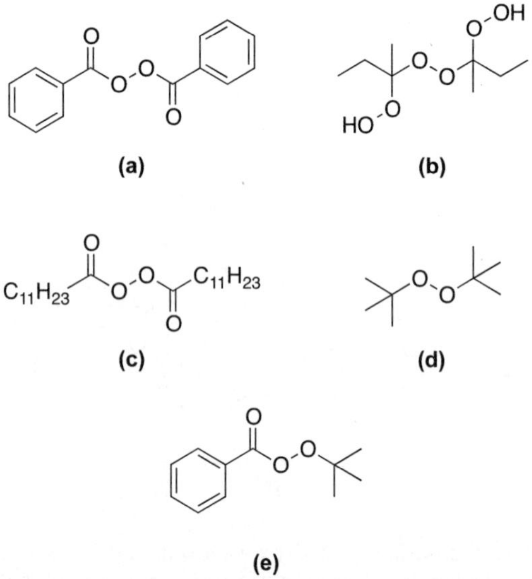

Figure 2.18 Peroxide initiators studied for acrylate-based self-healing: (a) benzoyl peroxide; (b) methyl ethyl ketone peroxide; (c) lauroyl peroxide; (d) *tert*-butyl peroxybenzoate; and (e) *tert*-butyl peroxide.

addition-fragmentation chain transfer polymerization (RAFT), have also been exploited as initiators for radical-based self-healing chemistries. These healing chemistries take advantage of the long half-lives of end-chain radicals for initiating polymerization of a released healing agent. Two related papers report that living radicals that are found within a polymer produced by ATRP or RAFT polymerization.[37,38] In the first report of this technique, a poly(methyl methacrylate) polymer was formed using an ATRP polymerization and an acrylate healing agent, glycidyl methacrylate (GMA), was encapsulated and dispersed in the monomer solution prior to polymerization. After the capsules were ruptured by a crack, the GMA polymerization was initiated by the 'living radicals' within the polymer.[37] Impressive healing efficiencies approaching 100% were reported. However, in this system, healing had to be carried out under an inert atmosphere, as exposure to air quickly deactivated near-surface radicals. Attempts to minimize this drawback included investigating the use of a bulk matrix produced by RAFT polymerization, in which the living radicals are more stable in air.[38] This healing system utilized the same microencapsulated healing agent, GMA, and also exhibited high healing efficiencies.

2.3.4 Isocyanate-Based Healing Systems

Isocyanate systems would seem to be ideal for self-healing applications as they exhibit fast reaction rates, frequently without any initiator or catalyst, utilizing environmental moisture as a cross-linker. Unfortunately, the very characteristics that make this system ideal are also the same characteristics that typically prevent isocyanates from being encapsulated. Two reports in the literature have described how this limitation can be successfully overcome to produce microcapsules containing highly reactive isocyanates. The first demonstration of liquid isocyanate encapsulation utilized an aliphatic isocyanate, IPDI, encapsulated using a prepolymer based upon TDI, an aromatic diisocyanate. By taking advantage of the large disparity in reaction rates between the aromatic and aliphatic diisocyanates, the liquid IPDI was successfully encapsulated within a polyurethane shell using interfacial polymerization.[14] However, in this case, reactivity of IPDI released from the microcapsule was demonstrated, but the system was not self-healing. Subsequently, Huang and Yang successfully adapted this encapsulation approach to produce microcapsules containing hexamethylene diisocyanate by using a methyldiphenyl diisocyanate prepolymer for the interfacial poly-merization.[15] This system did demonstrate the qualitative capability to self-heal.

2.3.5 Epoxy-Based Healing Systems

Epoxy-based healing systems were the first self-healing chemistries that were based on the incorporation of a secondary phase in a host material. Zako and Takano incorporated solid particles of a high temperature curing epoxy material within a room temperature curing epoxy composite.[39] After damage, heat, between 110 and 140 °C, was applied to melt the embedded particles, which then flowed into the crack plane and cured. However, the concept of healing effi-ciency had not gained widespread understanding at that time and so the results were presented as a simple comparison of healing and non-healing performance.

Since this first demonstration, several groups have reported healing based on the ring-opening polymerization of epoxy systems. A significant proportion of the research investigating the performance of hollow fiber-based self-healing materials has been performed using epoxy resin and amine curing agent-based healing systems.[18,40–45] These systems avoid the complexity and difficulty of encapsulating amine curing agents, as the hollow fiber encapsulation procedure is physical in nature. Fibers are embedded in a material and the healing agents are then infiltrated using a vacuum process. Most hollow fiber systems directly encapsulate commercial room temperature curing epoxies as healing agents. Healing efficiencies have been assessed by low-speed impact and can be very high, up to 91% recovery of flexural strength.[44]

The first report of successful microcapsule-based healing using epoxy chemistry was of that based on a latent hardener complex $CuBr_2(2\text{-MeIM})_4$.[46] To accomplish healing, an epoxy prepolymer was encapsulated within UF microcapsules and was then dispersed in an epoxy resin. The epoxy matrix was blended with the latent hardener complex prior to curing. This system required the application of heat, 130 °C, to initiate and complete the poly-merization of the released epoxy. Healing efficiencies were between 40% and 90%, depending on concentrations of the latent hardener and the micro-capsules. Other catalytic curing agents have been encapsulated and utilized as initiators for the polymerization of epoxy prepolymers. A microcapsule-based material with scandium triflate as the catalytic curing agent was recently demonstrated.[47] Amercaptan healing agent has also been encapsulated and used in self-healing polymers and composites and demonstrated healing at room temperature.[48]

Recently, amine-based hardeners have been successfully microencapsulated using a particle-stabilized reverse water-in-oil emulsion.[13] The encapsulation of amines has proven to be difficult as most useful amine-based hardeners are readily water-soluble and therefore not directly amenable to the encapsulation approaches described. Interfacial encapsulation was used to encapsulate an amine adduct and the emulsion was stabilized by using nanoparticles. Healing was demonstrated in a self-healing adhesive formulation, assessed by fracture and fatigue.[49]

2.3.6   "Click-Chemistry"-Based Healing Agents

Click chemistry is a philosophy rather than a specific chemical system, which requires a click reaction to have the features of high reactivity, specificity, stable products, and few or innocuous by-products.[50] These reactions are exemplified, most notably, the azide–alkyne Huisgen [3 + 2] cycloaddition process, shown in Scheme 2.4.

While no direct reports of successful self-healing utilizing this chemistry are available in the literature, there has been a successful encapsulation of a click-chemistry system, using the alkyne and azide healing agents shown in Figure 2.19.[51]

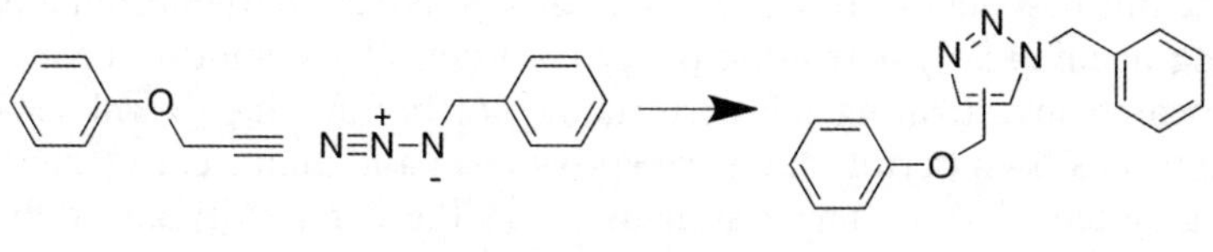

Scheme 2.4   Example of the Huisgen [3 + 2] cycloaddition reaction, frequently now referred to as a 'click' reaction.

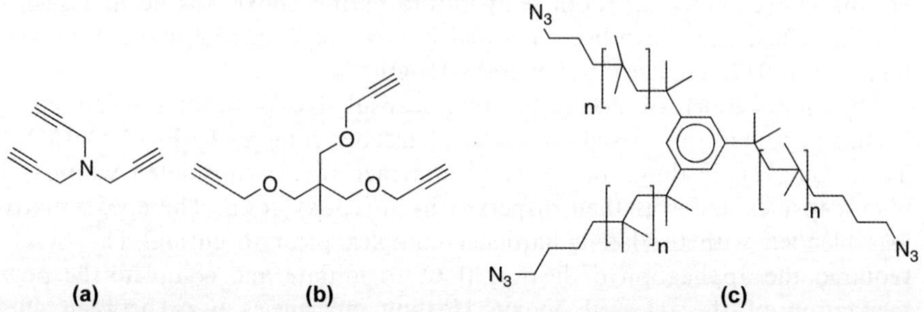

Figure 2.19   Healing agents for proposed azide–alkyne "click-chemistry"-based self-healing material.

These monomers were encapsulated in a UF shell using the same approach as that outlined by Brown and co-workers.[5] The reaction between the monomers shown in Figure 2.19 was catalyzed by a copper complex, $CuBr(PPh_3)_3$, embedded in the host matrix. To demonstrate reactivity, the microcapsules were mixed into a lightly cross-linked poly(isobutylene) matrix and then deformed to rupture the capsules. Reactivity was assessed by measuring the increase in modulus that was generated by the cross-linking of the rubbery polymer with the released capsule contents.

2.3.7 Solvent or Plasticizer-Based Healing Chemistries

Solvent-based healing chemistries provide an interesting alternative to the heterogeneous adhesion approaches discussed, and the first studies of crack and damage healing were performed using solvents.[1] However, these early studies required the use of higher temperatures or pressures and long timespans to fully recover material properties. Unlike the chemistries described previously—which involve adhesive bonding of a polymerized healing agent to a host polymer—solvent-based chemistries rely on promoting the local interdiffusion of the host polymer across the fracture surface. Several reports of successful self-healing have been published whereby a solvent,[24,52–54] or a combination of solvent and reactive agent,[35] was used as the healing agent. Solvents that have been successful in epoxy-based matrices include chlorobenzene, phenyl acetate, and ethylphenyl acetate.[20,35] Healing efficiencies are typically high in these approaches and can exceed 90%.

In addition to purely solvent-promoted healing, two hybrid solvent–latent functionality approaches have been reported. By mixing a small amount of epoxy resin material into a solvent and then encapsulating the resulting solution, healing efficiencies of 100% were achieved.[35] The proposed mechanism of healing in this system was a combination of solvent-promoted chain interdiffusion and the harvesting of latent amine functionality in the matrix. This proposed mechanism was confirmed by heat treating the self-healing material to ensure complete consumption of free amine within the cured matrix. After the heat treatment, the healing efficiency dropped significantly. As a further development of this approach, Meng *et al.* used encapsulated glycidyl methacrylate to take advantage of a combination of non-covalent interaction and covalent bonding of free amine within the host matrix.[24] This system also demonstrated 100% healing efficiencies.

2.3.8 Summary of Encapsulation-Based Self-Healing Materials

Since the initial introduction of ROMP-based healing chemistry, researchers have mined the polymer science literature for environmentally stable and encapsulatable healing chemistries. Many of the obvious chemistries have been implemented, but to date, a 'universal' encapsulation system that offers high healing efficiencies for all bulk matrix materials has yet to be achieved. Further developments include the potential of using microencapsulated nano-particles as an active healing chemistry.[55] The perfect combination of stability, reaction rate, and cost has not yet been achieved. Each of the chemistries described above is useful in specific circumstances and both the operating regime and matrix material will dictate selection when formulating a self-healing polymer or composite.

2.4   Performance and Testing of Microcapsule-Based Self-Healing Materials

As exemplified in the previous section, the performance testing of self-healing materials is typically based on the measurement of the recovery of some critical material parameter, *i.e.* the healing efficiency (see Chapter 1). In many of the encapsulation-based healing polymer systems, fracture toughness has been the most frequently measured material property.[2,56–58] However, modulus,[39] impact strength,[59,60] and tear strength[34] have also been utilized to assess healing efficiencies. In this section, the mechanical testing and performance of the various self-healing systems and encapsulation-based materials that have been developed will be discussed. For simplicity's sake, this section will be structured around the types of testing performed, for example fracture, impact, or tear.

As with all materials testing studies, appropriate and accurate control experiments are critical for the characterization of self-healing materials. This is especially true in self-healing materials, as the mechanical testing does not just represent a physical assessment, but also mechanism investigation (see Chapter 1). Most multi-component healing systems will have test regimen for each of the individual components, as well as the basic matrix material, to investigate the contribution of each phase and to eliminate any potential contribution of spontaneous self-healing of the matrix. This last point is particularly critical for elastomers or other polymers with high values of surface adhesion and relatively large free volume. In addition to the individual components, two specialized control tests, the injected heal and the self-activated heal, are usually performed. The injected heal uses pre-mixed healing agent to investigate the behavior of the healing chemistry when mixed in ideal stoichiometric quantities. The self-activated heal applies to multi-component systems and has the polymerization catalyst or initiator present in the host polymer, along with an unmixed healing agent, which is then injected onto the damage plane.

2.4.1   Fracture Testing

The earliest research in self-healing utilized a fracture-based material characterization approach. Recovery of fracture toughness was assessed using a double-cantilevered beam (DCB) geometry for a lightly cross-linked poly(butadiene) matrix.[1] While the standard DCB specimen is well characterized and understood from a fracture mechanics standpoint, this geometry has one major drawback: the need to know the location of the crack tip in order to calculate the fracture toughness. Determining the location of the effective crack tip in the initial test is a straightforward process; but during the healed test, finding the crack tip can become difficult, if not impossible. The solution to this practical issue arose from the work of Mostovoey, Crosley, and Ripling, who, in 1967, realized that the dependence of fracture energy on specimen compliance could be exploited to fabricate specimens that were crack length independent.[61] The relationship between specimen compliance $C(a)$, which is dependent on crack length, and fracture energy, G, is given by Equation (2), where a is the crack length and α is a parameter that depends on specimen width and failure load.

$$G = \alpha \, (dC(a)/da) \qquad (2)$$

For linear elastic materials, a valid assumption for many engineering polymers, the functional form of C can be determined using beam theory. Equation (3) is the case for the typical DCB specimen,[62] where P is the applied load, a is the crack length, E is the elastic modulus of the material, h is the height of the specimen, and B is the width of the specimen.

$$G = (12P^2 \, a^3)/(Eh^3 \, B) \qquad (3)$$

As can be seen from Equation (3), by choosing the appropriate height profile, h, or width profile, B, the fracture energy, and therefore fracture toughness, can be made constant with respect to crack length, a. Fracture toughness is related to fracture energy by the relationship given in Equation (4).

$$G = \begin{cases} \dfrac{1 - v^2}{E} K^2 : \text{Plane Strain} \\[2mm] \dfrac{K^2}{E} : \text{Plane Stress} \end{cases} \qquad (4)$$

Specimens that are constructed by keeping the width constant and varying the height profile are referred to as height-tapered double cantilevered beam (TDCB) specimens. When the height is constant and the width is tapered, these are referred to as width-tapered double cantilevered beam (WTDCB) specimens. The use of crack length independent specimens represented a critical step forward in the unambiguous characterization of the recovery of fracture properties. For the initial studies of the DCPD/Grubbs systems,[58] the specimen was a height-tapered fracture specimen, shown in Figure 2.20.

2.4.1.1 Quasi-Static Loading

Using a specimen with the geometry specified in Figure 2.20, healing efficiencies of greater than 90% were obtained for an optimized system using DCPD/Grubbs' catalyst.[58] Since this initial report, the TDCB specimen has been employed to characterize a variety of other healing systems. The GMA-based solvent/nucleophilic addition system described was also capable of recovering over 90% of the virgin fracture toughness.[24] Siloxane-based chemistries utilizing phase separation have also been investigated using TDCB specimens.

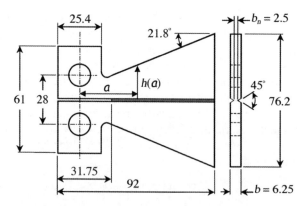

Figure 2.20   Height-tapered double cantilevered beam specimen (TDCB) all dimensions are in millimeters.
Reprinted with permission from reference [58].

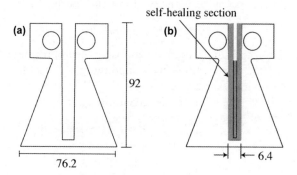

Figure 2.21   Localized self-healing TDCB, all dimensions are in millimeters.

The fracture strength recovery was low as a result of the mechanical strength mismatch between the healing agent-derived polymer and the epoxy material.[11] A mercaptan–epoxy system was also investigated using TDCB specimens, but exhibited variable healing efficiencies, varying from 50% to full recovery.[63] Solvent systems have also been characterized using this specimen geometry and have demonstrated healing efficiencies from 60% to 100% dependent on the particular solvent used.[20,35]

The investigation of the tungsten-catalyzed ROMP was carried out using a localized TDCB specimen, which allowed for a reduction in the amount of catalyst required for characterization of the healing response. In the localized TDCB shown in Figure 2.21, only the central section of the specimen is self-healing, reducing the amount of material required and the overall cost of the specimen. Recent work using a catalytic triflate-based polymerization initiator for a microencapsulated epoxy resin has also been studied with a localized specimen.[47]

One of the critical aspects of any self-healing system is that the inclusion of a self-healing functionality should not negatively impact the base performance of

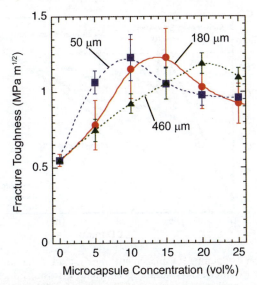

Figure 2.22   Effect of capsule size on virgin fracture toughness. Reprinted with permission from reference [64].

the material. If the basic performance of the material is significantly degraded in gaining the additional functionality, a pyrrhic victory has been won. In the case of microcapsule-based composites, the addition of capsules typically improves the fracture toughness of the matrix.[64] This effect is dependent on both the diameter of the microcapsules and the capsule concentration, shown in Figure 2.22 for an epoxy resin containing DCPD-filled UF microcapsules. Improvements in the fracture toughness are the result of the introduction of additional energy dissipation mechanisms, which manifest on the fracture surface as tails emanating from the trailing edges of microcapsules.

Fracture testing has also been utilized to investigate the recovery of interlaminar fracture toughness in woven composite materials.[56,57] Instead of a height-tapered specimen, a width-tapered test was used to assess the recovery of fracture strength. Healing was demonstrated for the composite, but clumping of the catalyst and microcapsules hampered the maximum healing efficiencies. Elevating the temperature during curing increased the healing efficiencies, but the process was no longer fully autonomic. Capsules were generally relegated to the resin-rich zones between the warp and weft tows of the woven fiber reinforcement, and the uneven distribution also impacted healing efficiencies.

The healing of adhesive joints has also been investigated using a fracture approach.[49] This approach used the DCPD/Grubbs healing system and an epoxy-based adhesive bonding two carbon-steel adherends. Test specimens were also width-tapered as for the composite fracture testing described previously. This self-healing adhesive demonstrated a healing efficiency of 56%, an example crack-opening displacement (COD) *versus* load plot for a self-healing adhesive system is shown in Figure 2.23.

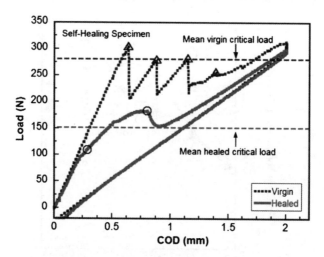

Figure 2.23 Crack Opening Displacement *versus* load behavior of an initial and healed interfacial fracture specimen for assessing adhesive bond strength. Reprinted with permission from reference [49].

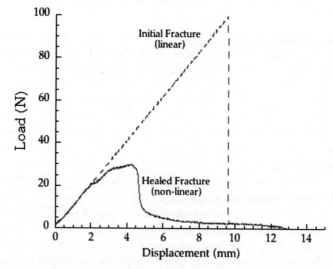

Figure 2.24 Comparison of a linear virgin fracture and a non-linear healed fracture. Image courtesy of Dr Joseph D. Rule.

Li *et al.* utilized a thermoplastic particle approach that required both heat and pressure to realize healing in an adhesive joint.[65] Complete recovery of the adhesive joint fracture strength was demonstrated, but the requirement of external intervention meant that the system was not truly self-healing in the manner defined in this chapter.

As discussed, several healing chemistries exhibit non-linear fracture behavior, shown in Figure 2.24. Non-linear fracture behavior can be the result of either a dissolved plasticizer, such as the wax surrounding a catalyst particle, or as the result of incomplete polymerization of the healing agent.

To test for this non-linear response, a fracture energy approach is typically adopted.[31] This method allows calculation of healing efficiency by comparison of the areas under each load displacement curve. Areas are found by numeric integration of the data and the healing efficiency is given by Equation (5), where U is the strain energy, given by the area under the curve, b_n is the width of the specimen, W is the distance from the loading line to the end of the specimen, and a_0 is the initial crack length.

$$\eta_G = \frac{\dfrac{U_{healed}}{b_n}\left(W - a_0^{healed}\right)}{\dfrac{U_{initial}}{b_n}\left(W - a_0^{initial}\right)} \tag{5}$$

Typically, the initial crack length is assumed to be approximately constant for the healed and unhealed specimens, and the parameters W and b_n are constant. Therefore comparison of the relative strain energies for the healed and unhealed specimens determines the healing efficiency.[31]

The diameters of the microcapsules incorporated are directly related to the size of the damage that can be healed. Rule and co-workers undertook a study to determine the relationship between capsule size and the crack volume that could be healed.[66] A simple relationship between capsule diameter, d_c, capsule mass fraction, ϕ, specimen density, ρ_s, and the mass of healing agent delivered, m, was derived [Equation (6)].

$$m = \rho_s \phi d_c \tag{6}$$

This relationship puts an upper limit on the degree of damage that a given capsule diameter can address. As shown in Figure 2.25, the peak load of fracture for a healed specimen, which can be considered as analogous to healing efficiency, is highly dependent on the volume of delivered healing agent. Thus the

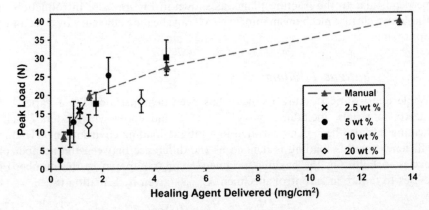

Figure 2.25 Peak load of healed specimen as a function of mass of delivered healing agent.
Reprinted with permission from reference [66].

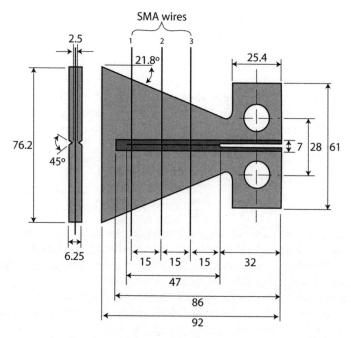

Figure 2.26   TDCB with active SMA crack closure.
Reprinted with permission from reference [67].

capsule diameter must be tailored to address the size of the expected damage. As might be expected, smaller capsules can be effective for small microcracking, but larger capsules are required to effectively repair large scale damage.

In an attempt to address the size limitation of the microcapsule approach, Kirkby *et al.* introduced shape memory alloy wires into the TDCB geometry in order to provide an active process to close cracks. These wires were embedded perpendicular to the fracture plane, as shown in Figure 2.26. Introduction of this active closure mechanism improved average healing efficiency by a factor of approximately 1.6.[67]

2.4.1.2   Fatigue Loading

While quasi-static fracture recovery has been demonstrated for a variety of matrix materials, including epoxy[47,54,35,58,68] and epoxy vinyl esters,[69] self-healing has also been demonstrated in a fatigue-loading environment. Healing efficiency for this loading is defined as the difference between the number of cycles to failure in the healing specimen N_{healed} compared to the number of cycles to failure in a control specimen $N_{control}$, given by Equation (8):

$$\eta = (N_{healed} - N_{control})/N_{control} \tag{7}$$

One of the most critical differences between the behavior of self-healing materials in quasi-static fracture testing and in fatigue is the basic mechanism of healing. For self-healing of quasi-static fractures, a single, adhesion-based mechanism is responsible for almost the entire recovery of fracture toughness. However, this is not the case for self-healing materials under fatigue loading, where several mechanisms operate to slow or arrest growing fatigue cracks.[70] The impact of each of the physical mechanisms initiated by the self-healing functionality on the effective stress intensity factor at the crack tip is shown in Figure 2.27. Prior to the rupture of any healing agent-filled microcapsules the crack tip, experiences the full applied crack tip stress intensity (indicated by region a). As the crack propagates and ruptures the capsules, liquid healing-agent is released onto the crack plane and a hydrodynamic shielding mechanism is introduced (indicated by region b). The mere presence of a fluid on the crack surfaces can significantly reduce the apparent crack tip stress intensity factor and lead to life extension when compared to a non-fluid filled crack tip.[71,72] As the healing-agent polymerizes, the full effect of the adhesive-based healing mechanism is realized (indicated by region c). If the adhesive bond is strong enough, or the applied loading low enough, the crack may arrest and stop propagating at this stage. However, if the loading is large enough, or the adhesive bond weak, the crack will begin to propagate again, and the wedge of polymerized healing agent will introduce a crack closure mechanism (indicated by region d). This closure will also act to slow the growth of fatigue cracks. Eventually, the crack may propagate beyond the healed zone into virgin material (indicated by region e), and the cycle will begin again.

Brown, White, and Sottos first investigated the behavior of the DCPD/ Grubbs system under tensile fatigue loading in a two-part study. Initially, only the behavior of control experiments was studied. Using a precatalyzed healing agent, this first paper elucidated the evolution of the mechanisms governing fatigue crack growth rate and arrest that were outlined in Figure 2.27. Subsequently, the performance of the fully *in situ* healing system was studied.[73] The fatigue performance of the material was highly dependent on loading conditions; however, the results did demonstrate that with appropriate

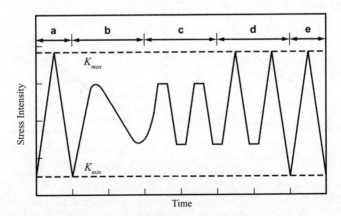

Figure 2.27 Load evolution in a self-healing fatigue test.

selection of loading, a fatigue crack could be slowed or even arrested as a result of the action of the self-healing functionality.

One of the critical findings of the investigation of fatigue behavior by Jones and co-workers was the realization of the importance of the interaction between the mechanical kinetics of the damage, the crack, and the chemical kinetics of the healing (the polymerization).[74]

To investigate this interplay, Jones and co-workers studied the behavior of the self-healing composite under high-cycle fatigue at lowered loads. In this loading regime, a self-healing material performed as if the polymer matrix possessed a fatigue threshold, as shown in Figure 2.28(a). For a higher-load, low-cycle fatigue loading, altering the chemical kinetics of the Grubbs/DCPD system that was being tested produced significant changes in performance, as shown in Figure 2.28(b). The change in chemical kinetics was achieved by varying the concentration and physical morphology of the Grubbs' catalyst present in the matrix material.[74] In Figure 2.28(b), curve 1 is a control specimen without catalyst, curve 2 contained 5 parts per hundred (pph) wax-encapsulated Grubbs' with an as-received morphology, curve 3 contained 5 pph wax-encapsulated Grubbs' that had been recrystallized, and curve 4 contained 5 pph wax-encapsulated Grubbs' that had been freeze-dried. The polymerization kinetics are influenced by the dissolution rate of the catalyst;[29] the use of the freeze-dried catalyst, which has fast dissolution kinetics, increased the polymerization rate of the released healing agent. As shown in Figure 2.28, increasing the rate of the chemical kinetics had a strong influence on the final fatigue life.

In addition to the variation of the chemical kinetics, Jones expanded on the investigation of the use of 'rest periods', initiated by Brown for improving the

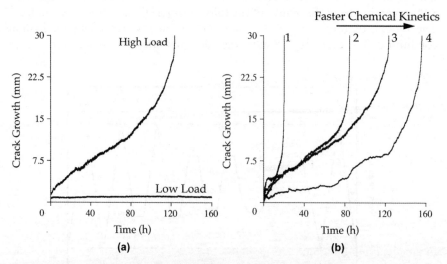

Figure 2.28   The impact of mechanical kinetics (a) and chemical kinetics, and (b) on the performance of a self-healing polymer under fatigue loading. Reprinted with permission from reference [74].

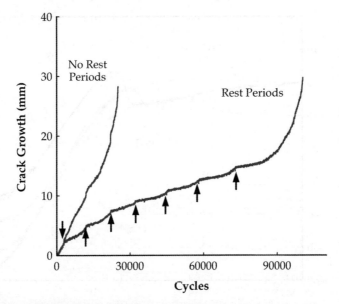

Figure 2.29 Impact of reset periods on the performance of a self-healing composite under fatigue loading. Arrows indicate rest periods.
Reprinted with permission from reference [74].

performance of the self-healing material.[74] These rest periods are periods of the test where the cyclic loading is paused and the chemical kinetics of polymerization are allowed to 'catch up' with the mechanical kinetics of damage. A comparison between a self-healing composite fatigued without rest periods and a composite with equally spaced rest periods is shown in Figure 2.29. The arrows in this figure represent points in the test where rest periods were introduced. The inclusion of rest periods has a huge impact on the overall lifespan of the material and reinforces the importance of the interplay between chemical and mechanical kinetics in self-healing materials.

The studies described were focused on the performance of a fatigue crack propagating in a bulk polymer. Fatigue has also been investigated in the performance of self-healing adhesive joints. Testing of these materials was performed using a similar width-tapered geometry as that used for the quasi-static fracture described. Under a stress intensity ratio of $R = 0.1$, the self-healing specimen demonstrated complete arrest of the propagating fatigue crack.

The torsion fatigue of elastomers has also been investigated, using the self-healing elastomer develop by Keller, White, and Sottos.[75] This mode III torsional fatigue test was chosen because of the combination of a closed crack-face that would reduce ejection of healing agent and acceptance as a model of fatigue loading for common elastomeric structures, such as the belt-edge region of tyres.[75]

The torsional stiffness curves shown in Figure 2.30 represent three control specimens and one fully *in situ* self-healing material. R-I is a non-healing

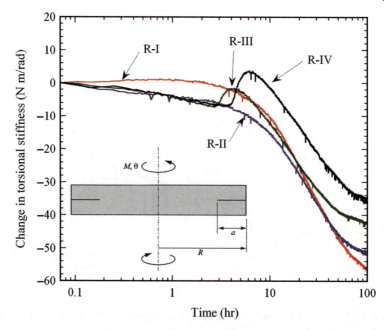

Figure 2.30   Stiffness evolution of the torsion fatigue of a self-healing PDMS elastomer. R-I is a non-healing control, R-II contained only resin capsules, R-III contained only cross-linker capsules, and R-IV is the fully *in situ* healing system.

control containing no microcapsules, R-II contained only resin capsules, R-III contained only cross-linker capsules, and R-IV is the fully *in situ* healing system. Both the *in situ* and cross-linker only specimens exhibited a stiffness increase at approximately the time required for the healing chemistry to gel (approximately 10 hours). The measured increase in stiffness at 10 hours represents the action of the polymerized healing agents, which slowed crack propagation. Post-testing measurements determined that the crack lengths in the two active samples were up to 24% shorter when compared to the non-healing control. This decrease in crack length was attributed to an artificial crack closure process similar to that found in the tensile fatigue loading, as represented by region d in Figure 2.27.

2.4.2   Impact/Ballistic Testing

There are two types of ballistic, or impact, testing that are performed on self-healing materials. Direct healing of impact damage involves a process whereby a self-healing material is formed into a Charpy or Izod impact specimen and then tested to failure. The second approach is to measure the recovery of a residual property, such as flexural strength, in a self-healing composite that has been damaged by impact or indentation.

Several research groups have characterized the self-healing of damage generated by impact on composite panels using the compression-after-impact (CAI) approach. In this test, a composite panel is damaged using either low-speed impact (indentation) or a drop weight tower. The panel is then loaded in a specially designed fixture and compressed until panel failure in order to assess a residual strength. Of the encapsulation systems that have been studied using this approach, the hollow fiber-based materials exhibit superior healing efficiencies compared to the microcapsule-based systems. Bond and co-workers investigated the performance of hollow-fiber-based graphite–epoxy composites.[44] These systems demonstrated excellent healing efficiencies, recovering between 70% and 90% of the undamaged CAI strength, depending on the force of the indentation. Microcapsule-based systems demonstrated significant recovery of compressive strength, up to 96% recovery, at low impact energies (*ca.* 18 J).[76] As the impact energy increased, the healing efficiency decreased. This effect was attributed to increasing crack volume requiring more healing agent than could be delivered by the embedded microcapsules. The effect of damage size on healing performance was described in Section 2.4.2.

Izod impact testing has been used extensively to characterize the healing of the living radical promoted materials.[37,38,77] In these studies, the Izod sample, a thin rectangle clamped in cantilevered beam geometry, is impacted and completely severed. The two halves of the sample are brought back into contact and allowed a healing period under a temperature controlled environment. In the case of the radical systems, the healing efficiencies were very high, but required healing in a desiccator under argon.[77] A related test, the Charpy impact test, was used to investigate the healing of an epoxy resin blended with a thermoplastic,[78] which also saw high healing efficiencies; however, elevated temperatures and pressure were required.

2.4.3 Tear Testing

Healing of elastomeric materials is frequently assessed by a tear specimen. Tear specimens come in a variety of geometries, but the so-called 'trouser-tear' test, shown schematically in Figure 2.31, has several advantages when compared to other test specimen configurations.

The primary advantage is the simple analysis of the specimen; healing efficiency is calculated by simply dividing the average tear force of the initial test by the average tear force of the healed test. Based on this tear test, the elastomer developed by Keller, White, and Sottos demonstrated high healing efficiencies and was capable of fully recovering the virgin tear strength under certain circumstances.[34] The full recovery of tear strength was surprising, and un-expected. These high levels of healing were attributed to local areas of strong adhesion between the healing chemistry-derived polymer and the surrounding matrix. As the tear approached a region with these properties, the tear would deviate into surrounding virgin material and proceed along a new path (see

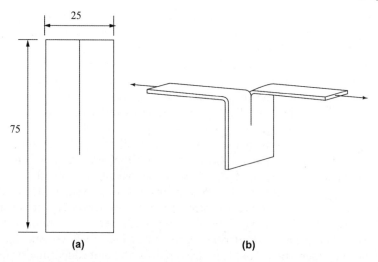

Figure 2.31   (a) Tear specimen geometry; and (b) schematic of a loaded tear specimen during testing.

Figure 2.32b). Furthermore, the addition of microcapsules improved tear properties for the virgin material, see Figure 2.32a.

2.4.4   Barrier Properties

Self-healing barriers have a long history in the paint and coatings industry, but self-healing in this context is not strictly the same property as that described in this chapter. Most 'self-healing' coatings take advantage of low T_g flow and creep to slowly level a scratch or other physical defect in the coating. Using the definition adopted in this chapter, a self-healing coating must possess a repair functionality that is triggered by mechanical damage, as with the cases described. The recovery of barrier properties is frequently assessed in a qualitative manner; such barrier properties include the recovery of corrosion resistance or in a 'go/no-go' recovery metric as in the case of pressure containing structures, such as fuel tanks. Quantitiative testing of self-healing barrier systems can be performed using linear voltammetry on barriers that are non-conductive.

2.4.4.1   Corrosion Barrier/Paint Healing

The first demonstration of autonomic recovery of barrier properties utilized a siloxane-based healing chemistry and investigated the self-repair of a coating on a steel substrate.[33] As shown in Figure 2.32, a self-healing coating that has been damaged using a scratch can recover complete barrier properties. The samples in Figure 2.32 were scribed using a commercial coating scriber, prior to resting for a 24 hour long healing period, and were then immersed in salt water. The recovery of barrier properties was found to be complete when the self-healing material was added to a commercial marine paint (Figure 2.33d). The

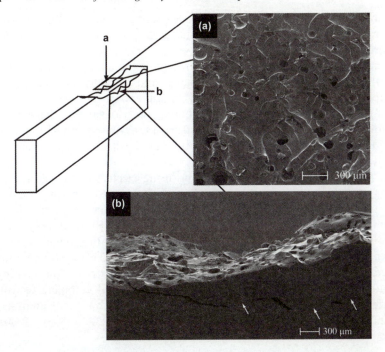

Figure 2.32 SEM images of a tear surface (a) and a deviated tear, and (b) for a self-healing PDMS elastomer.

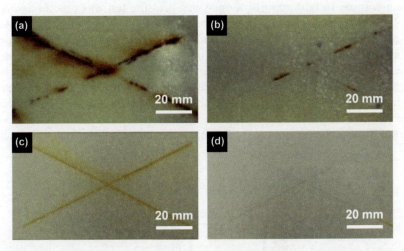

Figure 2.33 Recovery of corrosion resistance properties of a self-healing coating. Images (a) and (c) show control specimens with no self-healing functionality. Images (b) and (c) are self-healing coating materials. Images (a) and (b) are a custom made coating, and images (c) and (d) are commercial marine paints modified to become self-healing.
Reproduced with permission from reference [33].

incomplete recovery in a laboratory-made coating, Figure 2.33b, is a primarily the result of the poor scratch resistance of the base material (compare Figures 2.33a and 2.33c).

Since this first demonstration of film repair, there have been several additional reports in the literature. Huang and co-workers used a micro-encapsulated diisocyanate embedded in an epoxy film to repair scratch damage.[79] Since this system was based on isocyanate chemistry, this exploited 1-component healing chemistries in that it used ambient moisture to initiate and complete the polymerization of the healing agent released by the crack.

In the only demonstration of healing using electrospinning as the micro-encapsulation procedure, Park and Braun also healed scratch damage in a coating.[20] The polymer was a UV-cured polyurethane acrylate, which was required to prevent the poly(vinylpyrrolidone) (PVP) shell polymer from dissolving in the matrix. PVP is soluble in most coating polymers, such as polyurethanes and epoxies, but the rapid curing of the UV coating limited the time available for shell dissolution. The healing chemistry for this coating was an addition-cured PDMS. The material demonstrated complete recovery of barrier properties that was confirmed qualitatively, by visual analysis, and quantitatively, by using linear sweep voltammetry.

2.4.4.2   Pressure Barrier Healing

Two reports of restoring pressure-holding capacity have also been presented in the literature. In both reports, a one-sided pressure test was used to assess the healing of damage. The test set-up is shown schematically in Figure 2.34. A test is performed by first introducing damage to the membrane. The type of damage, either puncture or impact, is dictated by the base material of the specimen. Flexible, fabric-based membranes are typically punctured by a needle, whereas rigid polymer or composite membranes are impacted or indented. After damage, the membranes are allowed to heal and then placed in the pressure cell. Pressure is applied to one side of the membrane and the leakage pressure is measured on the opposite side. Healing efficiency is

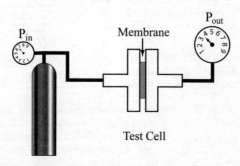

Figure 2.34   Schematic of the membrane–pressure support test cell.

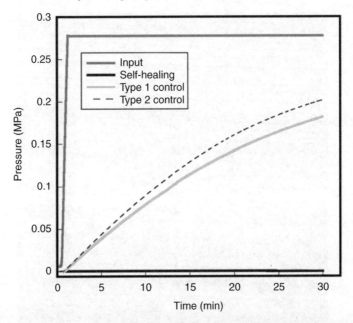

Figure 2.35 Pressure *versus* time traces for two non-healing controls (Type 1 and Type 2) and a fully self-healing material compared to the input pressure trace.
Reprinted with permission from reference [81].

calculated either as a 'leak/no leak' statistic based on a large number of samples or, potentially, as a reduction in gas flow rate as the result of the repair.

In the first report of an autonomic self-healing membrane, a flexible membrane based on siloxane chemistry was fabricated and demonstrated. This material consisted of a PDMS-based layer laminated between two high-performance barrier layers. Damage was introduced using a needle puncture to simulate pinhole damage as this system was intended for use in an inflatable space habitat.[80] Repair of this type of damage is of particular interest for the feasibility of composite-based fuel tanks. Moll and co-workers studied the recovery of pressure-holding capacity for a plain weave E-glass–epoxy composite. Square sample coupons were damaged with a sharp indenter, allowed a room temperature healing period, and then tested in a pressure cell similar to the one shown in Figure 2.33.[81] Results from this test are shown in Figure 2.35, which shows that the self-healing material was able to completely restore the pressure carrying capacity for the material when compared to the non-healing control specimens.

2.4.5 Interfacial Strength

In addition to the bulk material healing described, micro-debond specimens were utilized by Blaiszik and co-workers to investigate the healing of interfacial

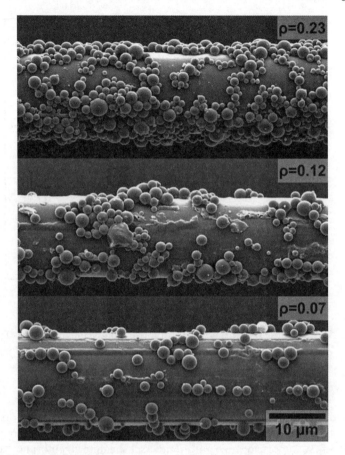

Figure 2.36   Nanocapsules adhered to the surface of a glass fiber.
Reprinted with permission from reference [82].

bonding in a model composite system.[82] The micro-debond specimen is a modified fiber pull-out test that attempts to measure the interfacial shear strength (IFSS) between an individual fiber matrix system.[83–85] A sample is manufactured by placing a small drop of epoxy or other matrix material on a single fiber, which is cured between two microscope slides. The fiber is then pulled out of the matrix drop. In the self-healing studies, the DCPD/Grubbs system[2] was deposited on the fiber/matrix interface as shown in Figure 2.36.

An alternative approach to the investigation of interfacial healing was presented by Sandana, Itaya, and Shindo.[86,87] Interfacial healing specimens were created by coating fibers with a slurry of DCPD microcapsules and Grubbs' catalyst in an epoxy resin that was starved of curing agent. The resulting fibers were then embedded within a tensile bar and loaded until failure. The separated halves were put back into contact, allowed to heal, and then loaded again until failure (Figure 2.37). Healing efficiencies of approximately 20% were achieved using this indirect test method.

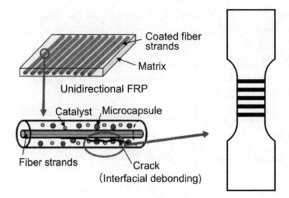

Figure 2.37   Tensile-based interfacial healing system.
Reprinted with permission from reference [87].

2.4.6   Conductivity

The restoration or repair of conductivity is a relatively new area for self-healing materials. To date, there have been only two reports of the autonomic restoration of conductivity; one system used a microencapsulated charge transfer salt, tetrathiafulvalene–tetracyanoquinodimethane (TTF-TCNQ), in a two-capsule formulation that formed a conductive salt when the two parts were mixed.[88] Healing was tested by fabricating a circuit with an induced flaw using sputter-coated gold on a glass slide. Capsules were crushed on the slide and the released contents formed a conductive path from one side to the other. The resistance of the healed circuit was measured and the resistance of the circuit was found to slowly decrease as time progressed. An alternative approach was also developed by the same group, using capsules filled with a solvent, hexyl acetate. For this self-healing system, conductive silver ink was used to create a circuit. A cut in a thin polymer coating over the conductive traces simultaneously broke the conductive path and released the solvent healing agent. The released solvent dissolved the polymer carrier in the silver paint, causing the paint to flow, and slowly restored conductivity to the initial level over the course of a week.[89]

2.5   Summary and Outlook

There has been rapid progress in the development of microcapsule-based self-healing materials in the 10 years since the first literature reports. Available healing chemistries have expanded and the microencapsulation procedures have become more varied and sophisticated. The general outlook for this class of materials is extremely promising in a wide variety of applications, especially in the structural composites industry. Commercialization efforts are underway and it is anticipated that a consumer or industrial-based product, most likely a coating or paint, will be introduced in the near future. Microcapsule-based healing is continuing to develop, with papers on new healing chemistries and formulations being published regularly. One of the most critical areas of research that is left is the incorporation of a function that would provide autonomic confirmation of healing. Ideally, any self-healing material would also come with the capability of informing an inspector that healing has been successfully accomplished.

References

1. R. P. Wool and K. M. O'Conner, *J. Appl. Phys.*, 1981, **52**, 5953.
2. S. R. White, N. R. Sottos, P. H. Geubelle, J. S. Moore, M. R. Kessler, S. R. Sriram, E. N. Brown and S. Viswanathan, *Nature*, 2001, **409**, 794.
3. S. Freitas, H. P. Merkle and B. Gander, *J. Controlled Release*, 2005, **102**, 313.
4. C. Thies, *Drugs Pharm. Sci.*, 1996, **73**, 1.
5. E. Brown, M. Kessler, N. Sottos and S. White, *J. Microencapsulation*, 2003, **20**, 719.
6. X. Liu, J. K. Le and M. R. Kessler, *Macromol. Res.*, 2011, **19**, 1056.
7. B. Blaiszik, N. Sottos and S. White, *Comp. Sci. Technol.*, 2008, **68**, 978.
8. D. S. Xiao, Y. C. Yuan, M. Z. Rong and M. Q. Zhang, *Polymer*, 2009, **50**, 560.
9. T. Nesterova, K. Dam-Johansen and S. Kiil, *Prog. Org. Coat.*, 2011, **70**, 345.
10. A. C. Jackson, J. A. Bartelt, K. Marczewski, N. R. Sottos and P. V. Braun, *Macromol. Rapid Commun.*, 2011, **32**, 82.
11. S. Cho, H. Andersson, S. White, N. Sottos and P. Braun, *Adv. Mater.*, 2006, **18**, 997.
12. M. Jacquemond, N. Jeckelmann, L. Ouali and O. P. Haefliger, *J. Appl. Polym. Sci.*, 2009, **114**, 3074.
13. D. A. McIlroy, B. J. Blaiszik, M. M. Caruso, S. R. White, J. S. Moore and N. R. Sottos, *Macromolecules*, 2010, **43**, 1855.
14. J. Yang, M. W. Keller, J. S. Moore, S. R. White and N. R. Sottos, *Macromolecules*, 2008, **41**, 9650.
15. M. Huang and J. Yang, *J. Mater. Chem.*, 2011, **21**, 11123.
16. M. M. Caruso, B. J. Blaiszik, H. Jin, S. R. Schelkopf, D. S. Stradley, N. R. Sottos, S. R. White and J. S. Moore, *ACS Appl. Mater. Interfaces*, 2011, **2**, 1195.
17. C. Dry, *Compos. Struct.*, 1996, **35**, 263.
18. J. W. C. Pang and I. P. Bond, *Comp. Sci. Technol.*, 2005, **65**, 1791.
19. S. D. Mookhoek, *Novel Routes to Liquid-Based Self-Healing Polymer Systems*, PhD Thesis, Technische Universiteit Delft, Delft, 2010.
20. J.-H. Park and P.V. Braun, *Adv. Mater.*, 2010, **22**, 496.
21. S. Sinha-Ray, D. D. Pelot, Z. P. Zhou, A. Rahman, X. F. Wu and A. L. Yarin, *J. Mater. Chem.*, 2012, **22**, 9138.
22. B. Blaiszik, M. Caruso, D. Mcilroy, J. S. Moore, S. White and N. Sottos, *Polymer*, 2009, **50**, 990.

23. X. Liu, J. K. Lee, S. H. Yoon and M. R. Kessler, *J. Appl. Polym. Sci.*, 2006, **101**, 1266.
24. L. M. Meng, Y. C. Yuan, M. Z. Rong and M. Q. Zhang, *J. Mater. Chem.*, 2010, **20**, 6030.
25. M. W. Keller and N. R. Sottos, *Exp. Mech.*, 2006, **46**, 725.
26. J. D. Rule and J. S. Moore, *Macromolecules*, 2002, **35**, 7878.
27. M. R. Kessler and S. R. White, *J. Polym. Sci., Part A: Polym. Chem.*, 2002, **40**, 2373.
28. X. Liu, X. Sheng, J. K. Lee, M. R. Kessler and J. S. Kim, *Comp. Sci. Technol.*, 2009, **69**, 2102.
29. A. S. Jones, J. D. Rule, J. S. Moore, S. R. White and N. R. Sottos, *Chem. Mater.*, 2006, **18**, 1312.
30. T. C. Mauldin and M. R. Kessler, *J. Mater. Chem.*, 2010, **20**, 4198.
31. J. Rule, E. Brown, N. Sottos, S. White and J. Moore, *Adv. Mater.*, 2005, **17**, 205.
32. J. M. Kamphaus, J. D. Rule, J. S. Moore, N. R. Sottos and S. R. White, *J. R. Soc. Interface.*, 2008, **5**, 95.
33. S. H. Cho, S. R. White and P. V. Braun, *Adv. Mater.*, 2009, **27**, 645.
34. M. W. Keller, S. R. White and N. R. Sottos, *Adv. Funct. Mater.*, 2007, **17**, 2399.
35. M. M. Caruso, B. J. Blaiszik, S. R. White, N. R. Sottos and J. S. Moore, *Adv. Funct. Mater.*, 2008, **18**, 1898.
36. G. O. Wilson, J. W. Henderson, M. M. Caruso, B. J. Blaiszik, P. J. Mcintire, N. R. Sottos, S. R. White and J. S. Moore, *J. Polym. Sci., Part A: Polym. Chem.*, 2010, **48**, 2698.
37. H. P. Wang, Y. C. Yuan, M. Z. Rong and M. Q. Zhang, *Macromolecules*, 2010, **43**, 595.
38. L. Yao, Y. C. Yuan, M. Z. Rong and M. Q. Zhang, *Polymer*, 2011, **52**, 3137.
39. M. Zako and N. Takano, *J. Intell. Mater. Syst. Str.*, 1999, **10**, 836.
40. I. P. Bond, R. S. Trask and H. R. Williams, *MRS Bull.*, 2008, **33**, 770.
41. J. Pang and I. Bond, *Compos. Part A-Appl. S.*, 2005, **36**, 183.
42. R. S. Trask and I. P. Bond, *Smart Mater. Struct.*, 2006, **15**, 704.
43. R. S. Trask, I. P. Bond and C. O. A. Semprimoschnig, 10th International Symposium on "Materials in a Space Environment", Collioure, France, June 2006.
44. G. Williams, R. Trask and I. Bond, *Compos. Part A-Appl. S.*, 2007, **38**, 1525.
45. H. R. Williams, R. S. Trask and I. P. Bond, *Smart Mater. Struct.*, 2007, **16**, 1198.
46. T. Yin, M. Rong, M. Zhang and G. Yang, *Comp. Sci. Technol.*, 2007, **67**, 201.
47. T. S. Coope, U. F. J. Mayer, D. F. Wass, R. S. Trask and I. P. Bond, *Adv. Funct. Mater.*, 2011, **21**, 4624.
48. C. Y. Yan, Z. R. Min, Q. Z. Ming, J. Chen, C. Y. Gui and M. L. Xue, *Macromolecules*, 2008, **41**, 5197.

49. H. Jin, G. M. Miller, N. R. Sottos and S. R. White, *Polymer*, 2011, **52**, 1628.

50. H. C. Kolb, M. G. Finn and K. B. Sharpless, *Angew. Chem., Int. Ed.*, 2001, **40**, 2004.

51. M. Gragert, M. Schunack and W. H. Binder, *Macromol. Rapid Commun.*, 2011, **32**, 419.

52. M. M. Caruso, D. A. Delafuante, V. Ho, N. R. Sottos, J. S. Moore and S. R. White, *Macromolecules*, 2007, **40**, 8830.

53. A. C. Jackson, J. A. Bartelt and P. V. Braun, *Adv. Funct. Mater.*, 2011, **21**, 4705.

54. S. D. Mookhoek, S. C. Mayo, A. E. Hughes, S. A. Furman, H. R. Fischer and Sybrand van der Zwaag, *Adv. Eng. Mater.*, 2010, **12**, 228.

55. G. V. Kolmakov, R. Revanur, R. Tangirala, T. Emrick, T. P. Russell, A. J. Crosby and A. C. Balazs, *ACS Nano*, 2010, **4**, 1115.

56. M. R. Kessler, N. R. Sottos and S. R. White, *Compos. Part A-Appl. S.*, 2003, **34**, 743.

57. M. R. Kessler and S. R. White, *Compos. Part A-Appl. S.*, 2001, **32**, 683.

58. E. N. Brown, N. R. Sottos and S. R. White, *Exp. Mech.*, 2002, **42**, 372.

59. Y. C. Yuan, Y. Ye, M. Z. Rong, H. Chen, J. Wu, M. Q. Zhang, S. X. Qin and G. C. Yang, *Smart Mater. Struct.*, 2011, **20**, 015024.

60. S. M. Bleay, C. B. Loader, V. J. Hawyes, L. Humberstone and P. T. Curtis, *Compos. Part A-Appl. S.*, 2001, **32**, 1767.

61. S. Mostovoy, P. B. Crosley and E. J. Ripling, *J. Mater.*, 1967, **2**, 661.

62. D.Broek, *Elementary Engineering Fracture Mechanics*, 3rd edn, Martinus Nijhoff Publishers, Boston, 1986.

63. Y. C. Yuan, M. Z. Rong, M. Q. Zhang and G. C. Yang, *Polymer*, 2009, **50**, 5771.

64. E. N. Brown, S. R. White and N. R. Sottos, *J. Mater. Sci.*, 2004, **39**, 1703.

65. G. Li, G. Ji and O. Zhenyu, *Int. J. Adhes. Adhes.*, 2012, **35**, 59.

66. J. D. Rule, N. R. Sottos and S. R. White, *Polymer*, 2007, **48**, 3520.

67. E. L. Kirkby, J. D. Rule, V. J. Michaud, N. R. Sottos, S. R. White and J.-A. E. Manson, *Adv. Funct. Mater.*, 2008, **18**, 2253.

68. C. L. Mangun, A. C. Mader, N. R. Sottos and S. R. White, *Polymer*, 2010, **51**, 4063.

69. G. O. Wilson, J. S. Moore, S. R. White, N. R. Sottos and H. M. Andersson, *Adv. Funct. Mater.*, 2008, **18**, 44.

70. E. Brown, S. White and N. Sottos, *Comp. Sci. Technol.*, 2005, **65**, 2466.

71. J. L. Tzou, C. H. Hsueh, A. G. Evans and R. O. Ritchie, *Acta Metall. Mater.*, 1985, **33**, 117.

72. J. L. Tzou, S. Suresh and R. O. Ritchie, *Acta Metall. Mater.*, 1985, **33**, 105.

73. E. Brown, S. White and N. Sottos, *Comp. Sci. Technol.*, 2005, **65**, 2474.

74. A. S. Jones, J. D. Rule, J. S. Moore, N. R. Sottos and S. R. White, *J. R. Soc. Interface*, 2007, **4**, 395.

75. M. W. Keller, S. R. White and N. R. Sottos, *Polymer*, 2008, **49**, 3136.

76. A. J. Patel, N. R. Sottos, E. D. Wetzel and S. R. White, *Compos. Part A-Appl. S.*, 2010, **41**, 360.
77. L. Yao, M. Z. Rong, M. Q. Zhang and Y. C. Yuan, *J. Mater. Chem.*, 2011, **21**, 9060.
78. S. A. Hayes, F. R. Jones, K. Marshiya and W. Zhang, *Compos. Part A-Appl. S.*, 2007, **38**, 1116.
79. G. C. Huang, J. K. Lee and M. R. Kessler, *Macromol. Mater. Eng.*, 2011, **296**, 965.
80. B. A. Beiermann, M. W. Keller and N. R. Sottos, *Smart Mater. Struct.*, 2009, **18**, 085001.
81. J. L. Moll, S. R. White and N. R. Sottos, *J. Compos. Mater.*, 2010, **44**, 2573.
82. B. J. Blaiszik, M. Baginska, S. R. White, and N. R. Sottos, "Autonomic Recovery of Fiber/Matrix Interfacial Bond Strength in a Model Composite," *Advanced Functional Materials*, 2010, **20**, 3547–3554.
83. S. Zhandarov, Y. Gorbatkina and E. Mader, *Comp. Sci. Technol.*, 2006, **66**, 2610.
84. S. Zhandarov and E. Mader, *J. Adhes. Sci. Technol.*, 2005, **19**, 817.
85. S. Zhandarov and E. Mader, *Comp. Sci. Technol.*, 2005, **65**, 149.
86. K. Sanada, N. Itaya and Y. Shindo, *Open Mech. Eng. J.*, 2008, **2**, 97.
87. K. Sanada, I. Yasuda and Y. Shindo, *Plast., Rubber Compos.*, 2007, **35**, 67.
88. S. A. Odom, M. M. Caruso, A. D. Finke, A. M. Prokup, J. A. Ritchey, J. H. Leonard, S. R. White, N. R. Sottos and J. S. Moore, *Adv. Funct. Mater.*, 2010, **20**, 1721.
89. S. A. Odom, S. Chayanupatkul, B. J. Blaiszik, O. Zhao, A. C. Jackson, P. V. Braun, N. R. Sottos, S. R. White and J. S. Moore, *Adv. Mater.*, 2012, **24**, 2578.

CHAPTER 3

Reversible Covalent Bond Formation as a Strategy for Healable Polymer Networks

CHRISTOPHER J. KLOXIN

Department of Materials Science and Engineering & Department
of Chemical and Biomolecular Engineering, University of Delaware,
150 Academy Street, Newark, DE 19716 USA
Email: cjk@udel.edu

3.1 Introduction

Polymeric materials are typically categorized as being either a thermoset or thermoplastic; this depends primarily on whether or not the material is covalently cross-linked. This convenient nomenclature often guides scientists and engineers to select one material over another for a particular application. While thermoplastics are capable of being remolded, remended, or recycled, the cross-links within a thermoset are permanently fixed upon fabrication. Thermosets are typified by their durability and strength; however, once cracked or fractured, these materials lose their structural integrity and must be replaced.

Recently, there has been renewed interest in covalent networks that are designed with reversible covalent linkages, termed covalent adaptable networks (CANs),[1] due to their capability to undergo bond shuffling and rearrangement. Such linkages impart unique properties that are an amalgamation of those associated with both thermosets and thermoplastics. Among these emergent properties is the ability to repair on a molecular level, designated here as

RSC Polymer Chemistry Series No. 5
Healable Polymer Systems
Edited by Wayne Hayes and Barnaby W Greenland
© The Royal Society of Chemistry 2013
Published by the Royal Society of Chemistry, www.rsc.org

'healing'. Reversible covalent cross-links enable a multifaceted approach to increasing the service lifetime of a material, from the subtle rearrangement of bonds leading to defect deletion, to macroscopic flow leading to larger scale repair of crazes and cracks. However, not all reversible covalent chemistry implementations lead to the same outcome of healing, and design principles are required for careful selection of reversible chemistries to enable covalent network adaptation and even healing.

This chapter will highlight the key design features of covalent adaptable networks and explain the underlying mechanisms that lead to a variety of healing phenomena. In doing so, several recent (and not so recent) chemistries will be highlighted. This is not to be taken as comprehensive review of this subject, in part because there are several excellent reviews on the matter,[2–5] but mostly because many of the new applications of reversible cross-links are deeply rooted in a vast number of both modern reactions and chemistries dating back several decades. Indeed, many of the recent reports on covalent adaptable networks are a reapplication of old materials and reactions. This is not to imply that these materials are trivial, but rather that there is renewed appreciation for the complex nature of these materials, which are well suited for many modern and specialized applications. Herein, recent advances in CANs will be placed within the wider context of the extensive literature concerning reversible covalent chemistries, outlining the key CAN design principles which will enable their widespread implementation in healing applications and provide a toolbox for the development of new CANs.

3.2   A Brief Historical Account

The idea that covalently cross-linked networks are capable of non-destructive creep has been well-known and extensively studied in vulcanized rubber since the 1930s and earlier.[6] In these materials, polymer chains are cross-linked by sulfur bridges, which are able to undergo scission and further cross-linking. A phenomenological description is utilized whereby an elastomer material is elongated to a prescribed strain and the material's equilibrium length is measured as a function of time. The change in material dimension is attributed to creep facilitated by the breaking and reforming of bonds.

Originally, the stress relaxation and creep observed in vulcanized rubbers was assumed to owe to strain induced polymer degradation; however, Green and Tobolsky[7] reasoned that the bonds must be continually breaking and reforming, based on the observation that the relaxation rates were independent of strain (for small extensions). Moreover, the bond breaking and reforming processes occur at equal rates and do not contribute to the stress for small deformations. This was a simple, and often overlooked, acknowledgement that these cross-linking reactions are at equilibrium. Nevertheless, the historic viewpoint of creep in cross-linked polymers was that it was considered to be a defect, and most efforts were focused on reducing this phenomenon or developing new materials that lacked this quality.

In the next several decades, there was a significant interest in covalent cross-linked materials that possessed reversibility. It was already well-known that many polymerizations possessed a ceiling temperature, which upon being reached, would reverse the material to a lower molecular weight (and even to its monomer constituents). Unfortunately, these high temperatures are often accompanied by irreversible side-reactions or decomposition, limiting this as a means of re-processing the polymer material. Indeed, the reduction of side-reactions remains a major challenge in the development of thermo-reversible covalent adaptable networks today.

In 1966, James Craven of the DuPont Company introduced a thermoreversible cross-linking material utilizing the reversibility of the [4 + 2] cycloaddition, or Diels–Alder reaction. In his 1969 patent, he wrote, "Polymer products should, for most uses, be insoluble in common solvents. For many purposes, such products should also have the ability to [be] post-formed... Heretofore, these properties were considered to be mutually exclusive."[8] This sentiment essentially captures the motivation at the time for developing processable covalent networks, as well the current development of new covalent adaptable networks.

In the last two decades, reversible, or dynamic covalent, chemistry (DCC)[3–5] has been utilized in several small molecule (*e.g.* dynamic combinatorial libraries)[3] and supramolecular chemistry[5,9] applications. Generally, these chemistries can be incorporated within multifunctional monomers that are used to form new covalent adaptable networks. The resulting properties of these networks are not only dependent on the kinetic and thermodynamic properties of the reversible covalent linkages, but also on the network design and connectivity. The subsequent section explores the various design elements that must be considered to create novel covalent adaptable networks for healing applications.

3.3   Covalent Adaptable Network (CAN) Design

The design of a CAN must take into account two main considerations: 1) network architecture, and 2) reversible chemistry type. The network structure is derived from the constituent monomers, which determine the location and distribution of reversible cross-links (*e.g.* Figure 3.1A & B). The type of reversible cross-link determines the method by which network strands rearrange on the microscopic scale (*e.g.* Figure 3.1C & D). The network architecture coupled with the reversible-cross-link type ultimately dictates how efficiently the material will heal or relax on a macroscopic scale.

3.3.1   Network Architecture

In general, there are two covalent adaptable network topologies: those created by reversible *monomers* (Figure 3.1A) and those created by reversible *macromers* (Figure 3.1B). The reversible monomer-based architecture tends to have a uniform distribution of reversible cross-links throughout the network backbone. In contrast, the reversible macromer-based architecture contains long, irreversible, contiguous chains that are cross-linked by reversible covalent bonds.

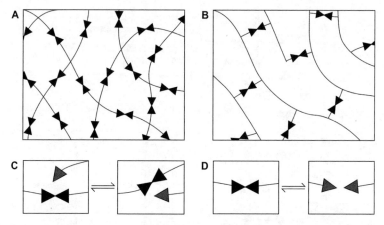

Figure 3.1   Different network architectures (A and B) and reversible cross-link types (C and D), where the bow-ties represent reversible cross-links. The networks shown in A (reversible monomer) and B (reversible macromer) are typically created using step- and chain-growth mechanisms respectively. The network rearrangement can occur by either reversible exchange (C) or addition reactions (D).

The formation of a reversible *monomer* network architecture typically follows a step-growth polymerization mechanism, which provides excellent control over the mechanical properties by simply changing the functionality (number of cross-linking functional groups per monomer). Moreover, the potential for enhanced concentration and uniform distribution of reversible cross-links makes this strategy highly desirable for self-healing materials. The reversible covalent cross-linking reaction can either be part of the network forming chemistry (Figure 3.2A) or latent within the backbone of the monomer (Figure 3.2B). In the former case, there is at least one reversible linkage between every junction point, or cross-link. Additionally, there are a number of suitable addition polymerization reactions, such as the disulfide formation (Section 3.4.3) and the furan-maleimide cycloaddition (Section 3.4.4). In the latter case, a large variety of reversible covalent linkages can be employed, so long as they are orthogonal (*i.e.* do not interfere) with the polymerization reaction.

Reversible *macromer* network architectures are readily distinguished by the presence of a long irreversible polymer chain that is capable of entanglement (*i.e.* the actual molecular weight exceeds the entanglement molecular weight). This criterion was appended to distinguish macromers from large molecular weight 'monomers'. The long irreversible chain can significantly affect the overall relaxation and adaptability of the polymer network, since even in the absence of cross-links, the polymer would have to diffuse *via* reptation. Thus the ultimate relaxation for macromer-based CANs is a balance between the timescale of reptation and lifetime of the cross-link (*i.e.* diffusion *versus* kinetics). Whereas this can complicate the relaxation process, the irreversible chains provide stability while the reversible cross-links enable the material to be remolded or remended.

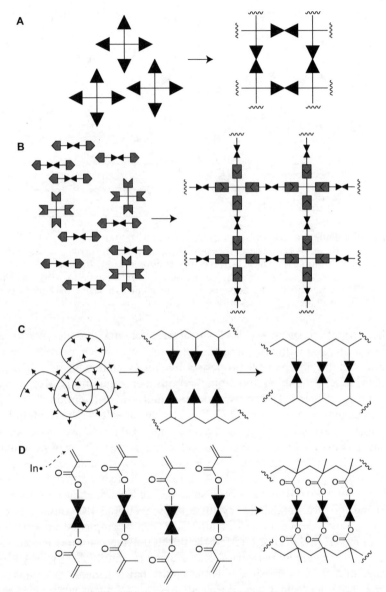

Figure 3.2   Select covalent adaptable network forming strategies. The dark grey triangles and 'bow-ties' represent a reversible covalent cross-link; the light grey irregular pentagon and forked square (B) represent non-reversible cross-links; and the wavy line represents a connection to the greater network. The network topologies produced in A and B are reversible *monomer* architectures, and those produced in C and D are reversible *macromer* architectures. The monomers used in A and C utilize the reversible linkages to form the network, whereas B and D possess reversible linkages in the macromer backbone, separate from the poly-merizable functional groups.

Reversible macromer networks are formed either by cross-linking pendant reversible functional groups (Figure 3.2C), or the chain polymerization of multifunctional groups flanking a reversible functional group (Figure 3.2D). In the former case, the polymer is synthesized using similar chemistries employed in the monomer strategy and has pendant functional groups capable of reversible addition reactions (see Figure 3.2A). As with reversible monomer-based networks, the latter case requires compatibility of the reversible linkage with the polymerizable functional group. Moreover, the polymerizable moiety must be difunctional, allowing two reactions, for example radical mediated polymerization of meth(acrylate) (as shown in Figure 3.2D) and ring-opening metathesis polymerization (ROMP) of norbornene.

3.3.2 Reversible Covalent Cross-link Type

The dynamic covalent cross-links that constitute a CAN are classified as being either a reversible *exchange* or a reversible *addition* reaction (Figures 3.1C and D respectively). In Section 3.4, a select group of chemistries that constitute these types of reactions are examined; exchange reactions preserve the network connectivity, while addition reactions break and reform and disrupt the network connectivity. Networks employing these two reaction types exhibit vastly different dynamics on the molecular scale, leading to significant differences in macroscopic behavior. Thus the network architecture and the reaction type must be considered concurrently in order to properly design a CAN for a particular application.

3.3.2.1 *Thermodynamic-* versus *Kinetic-Controlled Reversible Covalent Cross-links*

Reversible network chemistries are characterized as being either thermo-dynamically or kinetically controlled. While the ultimate conversion of any reaction is dependent upon temperature, the reaction rate can be insufficient to achieve equilibrium on practical timescales. Thermodynamically controlled covalent chemistry is often associated with DCC,[3–5] which is a group of reactions that are at equilibrium and readily susceptible to changes in thermodynamic variables (*i.e.* temperature, pressure, concentration, *etc.*); ideally, these reactions exhibit fast reaction kinetics, enabling them to rapidly respond to environmental changes. In contrast, kinetically controlled CANs are typically associated with a triggering event that initiates bond rearrangement. For example, polymer networks capable of undergoing radical-mediated bond exchange inherently exhibit no rearrangement until a radical is generated (*i.e.* by a photo-, thermo-, or redox initiator).

One critical difference between reversible covalent chemistry in small molecules and polymer networks is the phenomenon of vitrification. When the temperature is below the material's glass transition temperature (T_g), the underlying molecular dynamics responsible for rearrangement are kinetically frozen until the temperature is elevated above the T_g. Thus, when designing a material for a specific healing application, one must consider the state of the material at a given temperature. *A priori* predictions of the T_g based on molecular parameters are not well developed; however, decreasing the polymer network flexibility and increasing side-group bulkiness generally reduces the T_g. While vitrification can complicate a particular healing strategy, it also presents an opportunity to control or trigger healing in a thermodynamically-controlled cross-linked polymer (*i.e.* by elevating the temperature above the material's T_g). Materials that exploit glassy dynamics while also possessing exchangeable reversible covalent cross-links are termed vitrimers[10] and exhibit strong glass-forming behavior, similar to silica.[11,12]

3.3.2.2   *Triggering Thermoreversible Healing*

The nature of dynamic covalent cross-links, whether they are at equilibrium or triggered *via* an external stimulus, is an important consideration when tailoring the material for a specific application. Covalent adaptable networks respond to mechanical stress or material failure by either a *passive* or *active* mechanism. For those reversible chemistries that are at thermodynamic equilibrium, the network is continually rearranging, which can passively heal microfractures, blunt the initiation site of a crack, or simply reduce material defects that may have developed during fabrication. In contrast, active healing mechanisms are initiated in response to an external stimulus, such as light or mechanical stress. It should be noted that the latter stimulus (mechanical stress) is often referred to as self-healing, though all passive mechanisms would fit within this category.

One apparent limitation to *passive* thermoreversible CANs is the need to heat the material to accelerate the healing process. Typical demonstrations of thermoreversible network healing consist of cracking or fracturing the sample followed by heating the material in an oven. The healing efficiency of such materials is therefore highly dependent upon heat transport, which depends on the sample geometry and the heating conditions in the oven. Moreover, such strategies are not readily applied to many real-life applications, where the material may be fabricated on a large scale or be difficult to remove. In these cases, it would be advantageous to have a method to externally trigger the network rearrangement.

Light absorption, producing heat, can be used to trigger a thermoreversible reaction that could be incorporated into a thermoreversible network.[13,14] While the heat generated is reasonably controlled *via* absorber concentration and light intensity, the heat transport considerations still can be complicated by the sample geometry and environment. Another solution is to trigger internal heating of a composite CAN containing conducting particles by applying an alternating electromagnetic field, inducing a rapid alternating current in those 'susceptors' (*i.e.* Joule heating). Historically, this strategy has been used in thermoplastics for welding plastic together. While this approach allows greater penetration of the triggering electromagnetic field into the material, it ultimately suffers from many of the same heat transport complications as utilizing light. A related approach uses hysteresis heating of ferromagnetic nanoparticles. The induced current in nano-scale particles is typically insufficient to heat the material; however, the switching of magnetic domains within the ferromagnetic material results in a significant heating effect. Interestingly, ferromagnetic materials have a temperature ceiling—known as the Curie temperature—above which their magnetic domains become randomized and the material loses its magnetic character. Thus, by utilizing ferromagnetic nanoparticles in a thermoreversible CAN, the material will heat rapidly in the presence of an alternating electromagnetic field, but only up to the Curie temperature. This self-limiting heating mechanism can be used to ensure that the material is not over heated and reduces the probability of irreversible side reactions. Such a strategy has been demonstrated to remend a fractured thermoreversible Diels–Alder network embedded with chromium oxide nanoparticles at least 10 times without loss in mechanical strength.[15]

3.4 Reversible Covalent Cross-link Chemistries

There are numerous reversible covalent cross-link chemistries that are suitable for healing in covalent networks; in particular, DCC[3–5] has been adopted within supramolecular chemical and dynamic materials applications in the last two decades, providing a framework for new healing polymer networks. The discussions of covalent adaptable networks thus far have been restricted to reversible addition and exchange reactions since the production of a condensate inherently limits their reversibility and/or the stability of the network.[1,2] For example, the condensation reaction between multifunctional aldehyde and amine monomers produces a network cross-linked *via* an imine functional group (Schiff base) with water as a by-product (Figure 3.3A); thus the imine formation reaction is reversible in the presence of water. Nevertheless, the imine functional group will undergo bond exchange with pendent amine functional groups, leading to network adaptability (Figure 3.3B). Therefore, a wide range of dynamic covalent chemistries suitable for covalent network adaptability and healing are covered, with the caveat that the formation of a condensation product can present a major challenge when implementing these materials. Additionally, reversible reactions that exhibit possible irreversible side reactions should be carefully considered with regards to healing applications, as such reactions limit the number of healing cycles and decrease the lifetime of the material.

In the following sections, reversible chemistries that are suitable for healable covalent adaptable networks are highlighted. An attempt to list all such reactions would inevitably lead to omissions and, in such a rapidly growing field, will ultimately become obsolete. As the field of reversible chemistry is quite mature, it also would be difficult to identify any particular set of researchers associated with its development without overlooking classic studies from the early part of the last century. Nevertheless, there are several reviews outlining DCC[3–5,16] which could be implemented in CANs, or that directly address the utilization of reversible chemistry in covalent networks.[1,2]

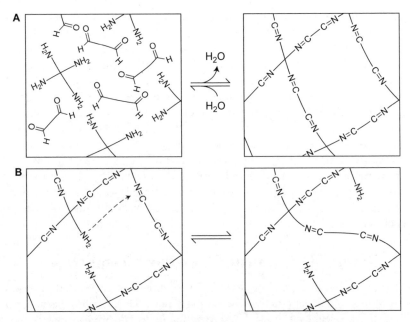

Figure 3.3 Formation (A) and exchange (B) reactions of an imine polymer network. The imine network is formed by the reaction of aldehyde and amine and requires the removal of water. When the network is formulated with excess amine, an exchange reaction with imine functional groups in the network backbone provides a route to network strand rearrangement.

The reactions outlined in Figure 3.4 are divided into reversible *exchange* and reversible *addition* reactions. Selection of reversible reaction type is one of the overarching considerations in designing a CAN for a particular healing application. Reactions (1) and (2) are listed under the 'trans-X' reaction type, which encompass most of the exchange type reactions, such as transimination, transetherification, transamidification, transesterification, *etc*. Addition–fragmentation reactions (3), which are often overlooked within the exchange reaction category, have received much attention since their inception into covalent networks less than a decade ago. Although, an argument could be made that the reversible disulfide bond (see reaction 4), which is also capable of exchange, is the first implementation of addition–fragmentation. The last reaction listed is the Diels–Alder cycloaddition (5), which is perhaps the 'cream of the crop' of thermoreversible bonds and also is considered one of the original 'click' reactions.[17] Metal–organic healing materials are not covered herein as they are not, or are only loosely, covalent in nature; but they are an important new class of materials that can have healing characteristics (see Section 4.5). The final section briefly describes several reversible reactions that do not strictly fit within the reaction types shown in Figure 3.4; these include: ionene, isocyanate–imidazole, olefin metathesis, nitroxide-mediated and other cyclo-additions reactions. The application of these disparate chemistries to CANs has

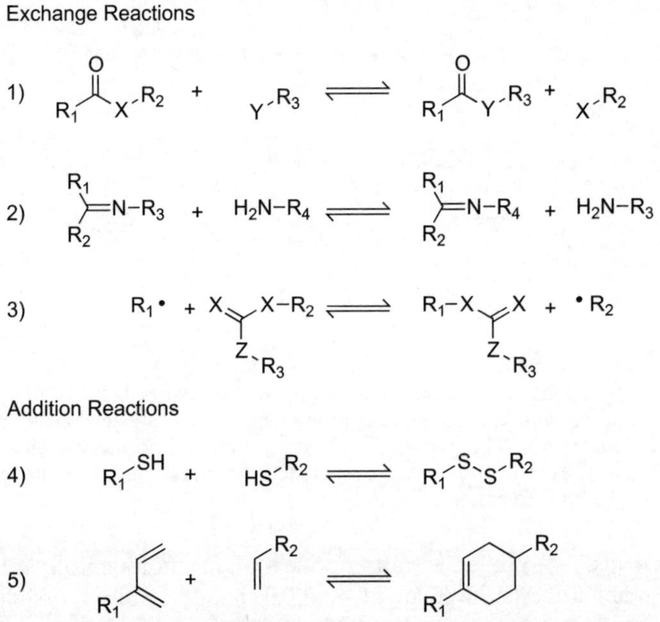

Exchange Reactions

1)

2)

3)

Addition Reactions

4)

5)

Figure 3.4 A selection of dynamic covalent chemistries for reversible cross-links in covalent adaptable networks: 1) transacylation; 2) imine exchange with amine; 3) reversible addition–fragmentation; 4) disulfide formation; and 5) the Diels–Alder cycloaddition.

been limited, either by detrimental irreversible side reactions, low efficiency, or that the methodologies have simply been underdeveloped. Nevertheless, as healing applications change, novel material implementations arise, or new reaction discoveries are made; any of these reaction types may be made practical as a reversible cross-link in covalent networks.

3.4.1 Trans-X Reactions

There is a broad class of non-radical exchange reactions, termed here as 'trans-X' reactions (*e.g.* transimination, transetherification, transamidification, transesterification, *etc.*). These preserve the overall number of cross-links while the network strands undergo connectivity rearrangement. In this section, we will outline several trans-X reactions that have recently been examined in CAN healing applications. By no means is this an exhaustive list, but is rather illustrative of well-known reversible chemical reactions that have been incorporated into covalent networks.

3.4.1.1 Transesterification

Transesterification reactions have been utilized in polymer science for several decades and have been extensively studied as a mechanism for relaxation in

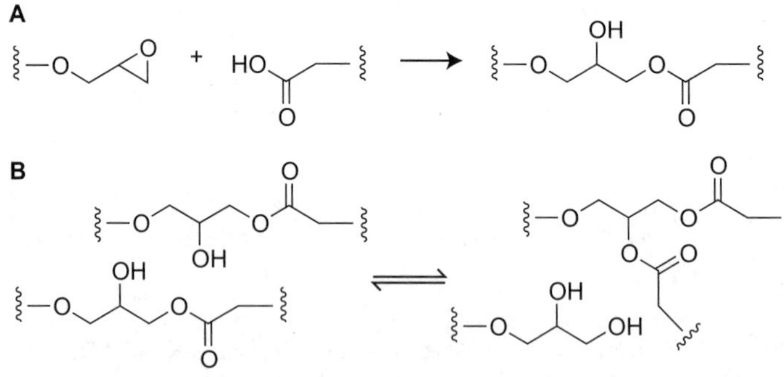

Figure 3.5 Epoxy-carboxylic acid network formation (A) and rearrangement (B) reactions. The formation of the polymer network between multifunctional epoxy and carboxylic acid monomers results in ester and hydroxyl bond formation. The hydroxyl and ester functional groups are able to undergo transesterification exchange reactions resulting in polymer network rearrangement.

polymer blends.[18–21] While the transesterification reaction typically is thought of being between an ester and an alcohol, exchange is also possible between carboxylic acids or other esters. Recent attention has been given to the catalyzed transesterification reaction in covalent networks, which provides a mechanism for rearrangement and self-healing. Specifically, Montarnal *et al.*[11] have demonstrated an epoxy-based covalent network that exhibits significant bond rearrangement *via* transesterification, leading to remolding and healing of this typically intractable material. Utilizing a cross-linking reaction between multifunctional epoxy and carboxylic acid monomers (Figure 3.5A), the resultant network possessed an ester functional group in the polymer backbone and a pendant hydroxyl functional group (Figure 3.5B). In this work, they demonstrated two different network properties simply by using different carboxylic acid comonomers; a diglycidyl ether bisphenol A produced a rubbery network with modulus $\sim 4\,MPa$ when mixed with multifunctional fatty acids, and a glassy network with a modulus of $\sim 2\,GPa$ when mixed with glutaric acid. As with many of these trans-X reactions, the use of a catalyst significantly increases the reversible exchange rates and is a critical component for many healing applications. Montarnal *et al.* used a metal catalyst (zinc acetate) that accelerated the kinetics by three orders of magnitude at $100\,°C$ relative to ambient, enabling thermoplastic-like processing at elevated temperatures and thermoset-like stability at ambient temperatures. Furthermore, the control over the reaction kinetics *via* the use of a catalyst provides a method to precisely tune the network dynamics for an application-specific timescale.

3.4.1.2 Trans-'siloxanification'

Siloxane-based polymer networks have recently been identified to possess great potential for healing covalent network applications. Zheng and McCarthy

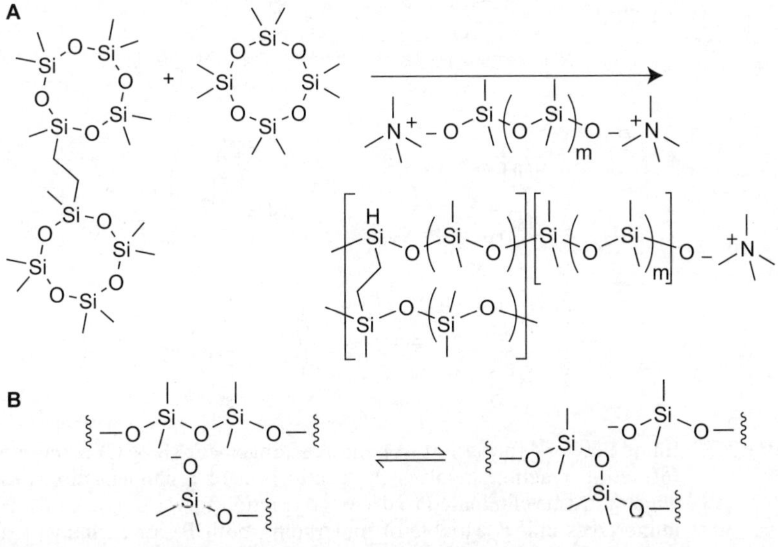

Figure 3.6 Siloxane network formation (A) and rearrangement (B) reactions. Mono- and di-functional octamethylcyclotetrasiloxane is ring-opened using bis(tetramethylammonium)oligodimethylsiloxanediolate as a ring-opening, anionic initiator. The living characteristic of the anionic end groups facilitates rapid rearrangement of the siloxane backbone.

revisited this network chemistry and highlighted the origin as being rooted in the 1950s literature.[22,23] In this study, Zheng and McCarthy formed a cross-linked network using a ring-opening difunctional cyclooctylsilane and controlled the mechanical properties by adding monofunctional cyclo-octylsilane as a chain extender (Figure 3.6A). The use of an anionic initiator resulted in a living polymerization that not only triggers the network formation, but also facilitates the rearrangement of the siloxane backbone (Figure 3.6B). In agreement with earlier results, Zheng and McCarthy showed that the ammonium-based anionic initiator is a critical component of the rearrangement mechanism.[24,25] Specifically, if the samples were heated above 150 °C, the chain ends decatalyze by alkyl transfer from the quarternary amine to the siloxane anionic end-group.

3.4.1.3   Transimination Reaction

As mentioned previously (see Section 3.4), imine cross-linking is a reaction typically performed between multifunctional aldehyde and amine monomers and requires the removal of water in order to obtain high conversions (Figure 3.7A). An imine polymer network is capable of undergoing transimi-nation in the presence of amines, leading to network strand reorganization (Figure 3.7B).[26] Furthermore, the imine bond is capable of undergoing a metathesis reaction with nearby imine network strands (Figure 3.7C).[27] While

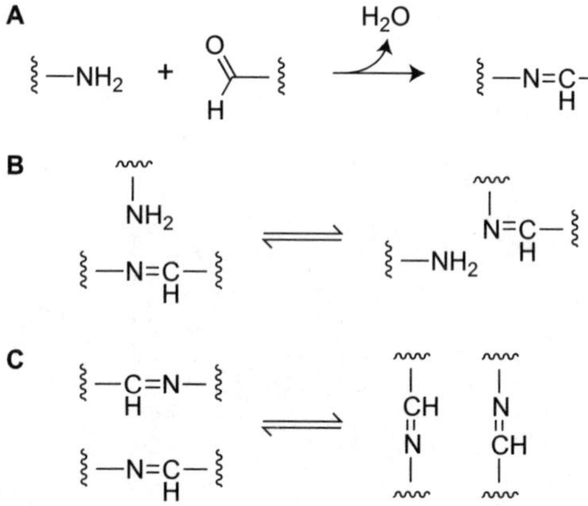

A $-NH_2$ + (aldehyde) $\xrightarrow{H_2O}$ $-N=C-$ with H

B $-N=C-$ with NH₂ $\rightleftharpoons$ $-NH_2$ with $N=C-$

C $-C=N-$, $-N=C-$ $\rightleftharpoons$ CH=N , N=CH

Figure 3.7 Imine network formation (A) and rearrangement (B & C) reactions. The formation reaction involves the production of a condensation product, which must be eliminated to drive the reaction forward. Once formed, the imine cross-link is capable of undergoing both B) an exchange reaction with amine, or C) metathesis reaction with another polymer strand containing an imine functional group.

several researchers have utilized the imine-based polymer network to create covalent organic frameworks,[28–30] the implementation as a healing material has yet to be fully realized, perhaps partly due to its water sensitivity. Nevertheless, the utilization of the imine linkages in dynamic covalent chemistry[27,31] and dynamic combinatorial libraries[3,32] suggests that this is a promising chemistry for healing materials applications.[4]

3.4.2 Radical-Mediated Addition–Fragmentation Reactions

The implementation of radical-mediated addition–fragmentation chemistry[33] within CANs is a fairly recent development.[34] This most likely owes to the fact that reversible addition–fragmentation chain transfer (RAFT) is typically identified with the synthesis of low polydispersity linear polymers or block copolymers. Nevertheless, it is an excellent method for triggering network rearrangement on demand, using light *via* a photo-radical generator (*i.e.* a photoinitiator) or by direct cleavage of the network strands.

The concept of radical-mediated reversible addition–fragmentation was recognized several decades ago *via* the intramolecular homolytic substitution (S_H2') reaction of an allylic functional group.[35–38] The first utilization of this mechanism to control polymerization was in the 1980s;[39–41] however, the full potential of the RAFT strategy to control polymerization polydispersity was not realized until the late 1990s.[42] A requirement of the RAFT strategy is that the reactants (before addition) and the products (after fragmentation) are the identical in all respects except for possibly molecular weight (*i.e.* RAFT is a degenerate chain transfer process; see reaction 3 in Figure 3.4). Therefore, to properly implement the RAFT-type strategy in covalent adaptable networks, one must consider the type of radicals generated. It should be noted that non-degenerative chain transfer processes may also be an effective strategy for network rearrangement, but will ultimately limit the extent of chain transfer events.

The first demonstration of addition–fragmentation chain transfer in a covalently cross-linked network utilized an allyl sulfide functional group within the network strands.[34] In this study, Scott *et al.* incorporated an allyl sulfide ring-opening monomer into a thiol–ene network formed by photopolymerization. The allyl sulfide ring-opening monomer, which was previously used to form linear polymers,[43–45] acts as chain extender in a thiol–ene step-growth polymerization (Figure 3.8). Utilizing latent photoinitiator, left over from the network formation step, Scott and co-workers were able to trigger addition–fragmentation within the polymer network backbone, yielding photoinduced plasticity. Importantly, they showed that the material's properties, specifically the elastic modulus, were statistically identical before and after the network rearrangement, demonstrating that the photoplasticity was not attributed to photodegradation.

The allyl sulfide functional group is well suited for radical-mediated thiol–ene polymerization owing to the degenerate chain transfer mechanism with which it reacts with thiyl radicals.[39,46] In an ideal thiol–ene polymerization, the carbon radical intermediate does not react with other vinyl functional groups (*i.e.* no homopolymerization), but rather abstracts a hydrogen from the thiol. Exploiting this concept, Bowman and co-workers demonstrated that by utilizing linear monomers having an allyl sulfide backbone, such a network rearrangement strategy has the additional benefit of dramatically reducing the polymerization-induced shrinkage stress.[46] They further demonstrated that a non-degenerative scheme (*i.e.* carbon radical addition to the allyl sulfide) can still significantly reduce polymerization stress, even in materials that exhibit glassy dynamics.[47–50] Finally, they demonstrated other applications for polymer networks containing allyl sulfides; one such example is the mechanophotopatterning that enables a photo-induced surface topology and shape changes to a covalent network after network fabrication.[51] It should be noted that all the cross-linked polymers described herein not only have the capacity to heal, but also to undergo a change in the material's equilibrium shape.

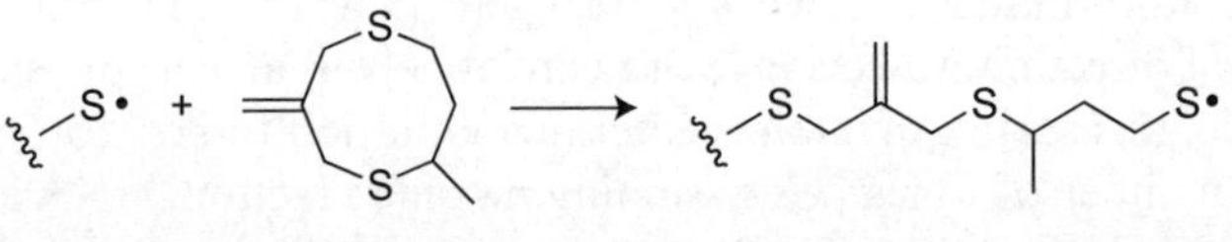

Figure 3.8 Thiyl radical (generated within the thiol–ene cycle) reaction with 2-methyl-7-methylene-1,5-dithiacyclooctane (MDTO), which ring-opens, inserting an allyl sulfide functional group and regenerating the thiyl radical.

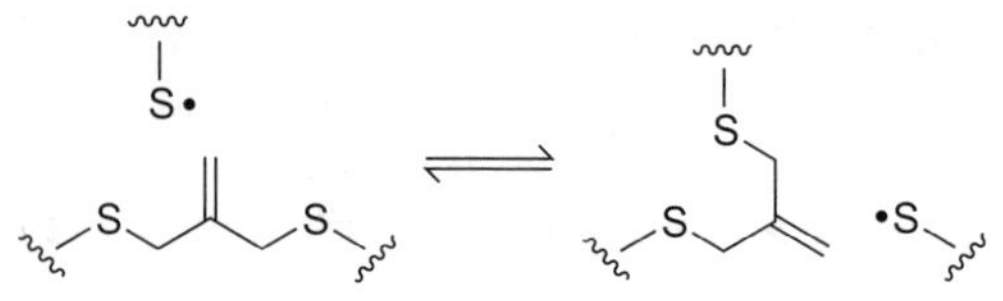

Figure 3.9   Thiyl radical reaction with allyl sulfide (addition–fragmentation), which preserves the overall chemistry (*i.e.* is non-degenerate) but results in a change in bond connectivity.

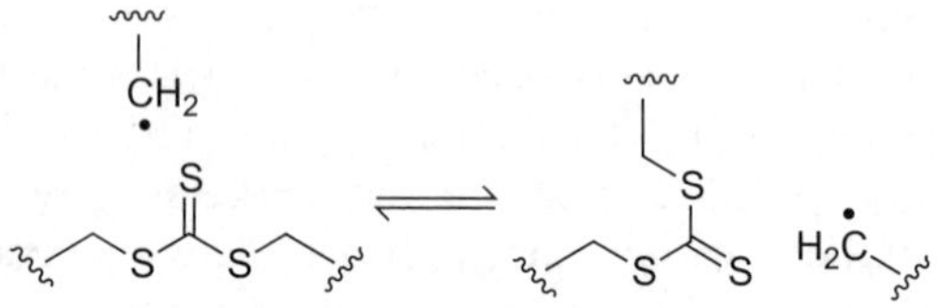

Figure 3.10   Carbon radical reaction with a trithiocarbonate, which preserves the overall chemistry whilst enabling the rearrangement of bonds (see Figure 3.9).

The addition–fragmentation chain transfer concept was further expounded using a conventional RAFT chain transfer agent (CTA, see Figure 3.10). Matyjaszewski and co-workers utilized a trithiocarbonate (TTC) functional group in the backbone of a di(methylmethacrylate)-based monomer to produce, *via* a chain-growth mechanism, a polymer network having RAFT functional groups in the cross-link (*i.e.* a reversible macromer architecture, Figure 3.1B).[52,53] The TTC functional group has been used as a RAFT agent in a number of linear polymer fabrication strategies[54–58] and is capable of degenerate chain transfer with carbon-centered radicals. In the first demonstration of healing *via* this approach, a polymer gel was formulated using methyl methacrylate and TTC dimethacrylate monomers in the presence of AIBN and swollen in anisole. Deoxygentated styrene and CuBr/ pentamethyldiethylenetriamine were introduced to three discrete pieces of this fabricated gel and the system heated to 60 °C for four hours. The styrene was able to add across the TTC functional group, mending the discrete pieces into a single cross-linked gel.

Amamoto *et al.* have further demonstrated the reshuffling of the trithiocarbonate functional group by simply irradiating the material with UV light.[52,59] The inclusion of a photo-radical generator is not necessary for triggering the photo-induced network rearrangement as the TTC functional group is capable of direct photo-cleavage of the relatively weak carbon–sulfur bond.[60] In principle, this strategy enables material adaptation to be photo-triggered an unlimited number of times (*i.e.* assuming no side reactions). Nevertheless, the wavelength and intensity necessary to generate a sufficient number of radicals to trigger the addition–fragmentation cascade is limiting and most likely causes other irreversible reactions. To address this, Amamoto *et al.* utilized a thiuram disulfide cross-linked material, which is readily cleaved using wavelengths into

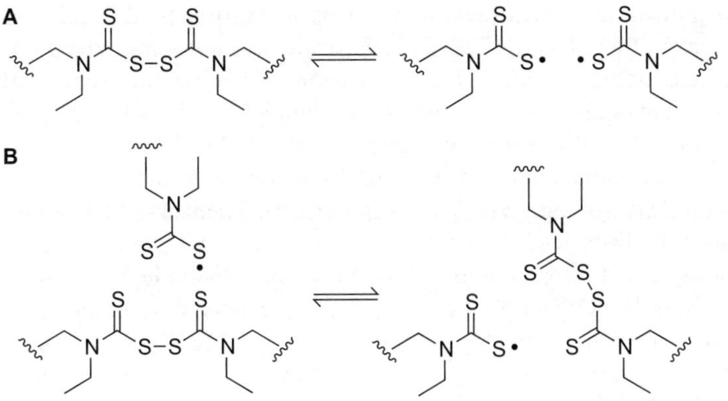

Figure 3.11   Formation of thiuryl radicals *via* homolytic cleavage (A) and subsequent addition–fragmentation of a thiuriam disulfide functional group.

the visible spectrum (Figure 3.11A).[61] Upon irradiation, the relatively weak disulfide bond is cleaved, producing two stable (*i.e.* long-lived) thiuryl radicals. Superficially, this process appears to be an addition reaction rather than an exchange reaction. However, Amamoto *et al.* argue that the thiuryl radical is able to undergo subsequent addition–fragmentation with thiuram disulfide functional groups, thereby undergoing a cascading bond-exchange process (Figure 3.11B). Moreover, the network was constructed using isocyanate-alcohol chemistry, producing a network *via* a step-growth mechanism, with a reversible monomer architecture. The authors utilized this attribute to control the number of TTC functional groups per cross-link, maximizing the addition–fragmentation network rearrangement effect.

The utilization of addition–fragmentation chain transfer in covalent adaptable networks is still in its infancy. Beyond the allyl sulfide and trithiocabonate functional groups, there are a large number of RAFT chain transfer agents that are suitable for inclusion in the backbone of the polymer network.[62–65] There is no requirement for the addition–fragmentation cross-link to be symmetrical. In general, the Z and R groups of the base RAFT chain transfer agent can be appended with polymerizable functional groups (*e.g.* Figure 3.4, reaction 3). Moreover, the selection of the Z functional group allows precise control over the reactivity of the addition–fragmentation cross-link.

3.4.3   Disulfides

Reversible disulfide formation in covalent networks is the prototypical CAN cross-linking reaction. Sulfur vulcanization is one of the earliest commercial successes in polymer networks, and the reversibility of sulfur-cross-linked materials has been extensively studied. As mentioned previously, the general concept of CANs is rooted in the seminal work of Tobolsky and co-workers, who examined the phenomenon of stress relaxation in disulfide (and tetra-sulfide) cross-linked materials.[7,66,67] Historically, reversible cross-linking was largely considered to be a negative attribute. Today, this same attribute in disulfide cross-linked networks is being exploited in self-healing applications.[68–71] While the dynamic nature of the disulfide bond has been more broadly implemented in small molecule or functional tether applications (*e.g.* the release of bioactive molecules in drug delivery,[72–75] or in dynamic combinatorial libraries[76]), such concepts are readily extended to impart a range of responsiveness into the network *via* reversible disulfide cross-links.

The disulfide bond is an extremely versatile reversible linkage, capable of undergoing homolytic breaking and reforming, as well as exchange reaction with ether thiolate anions[77–79] or thiyl radicals.[66,70] The formation of the thiolate from a thiol is dependent upon its pK_a, and thus the network rear-rangement is controlled by the pH of the surrounding media (for a pH responsive material). The thiyl radical is typically generated *via* radical abstraction of the hydrogen, which is facilitated by a photo-, thermo- or redox-initiator. Moreover, the formation of disulfide cross-linked polymers is readily achieved by the oxidation of thiol monomers or macromers (*e.g.* using I_2 or $FeCl_3$); the polymer can also be reduced to its constituent parts (Figure 3.12A).[80]

While disulfide cross-linked polymers have been utilized for over a century, the deliberate incorporation of disulfide cross-links for reversible network formation has only been explored in the last couple of decades. In contrast to the uncontrolled incorporation of sulfur *via* vulcanization, polymer chemists have systematically incorporated disulfide cross-linkers. Chujo *et al.* demon-strated the reversible oxidation and reduction of disulfide cross-linked poly(*N*-acetylethylenimine) by utilizing a cross-linker that contains within it a disulfide functional group. More recently, Canadell *et al.* demonstrated healing in a disulfide network fabricated with a multifunctional thiol and difunctional epoxy and containing a disulfide in the monomer backbone (*e.g.* Figure 3.2B).[71] In this work, they utilized the ability of the disulfide linkage to undergo disulfide-disulfide exchange reaction. The bond rearrangement mechanism involves the breaking of a disulfide bond and a subsequent thiolate reaction with another disulfide bond (Figure 3.12B). Others have demonstrated that the direct disulfide-disulfide interchange (*i.e.* metathesis) can be achieved in small

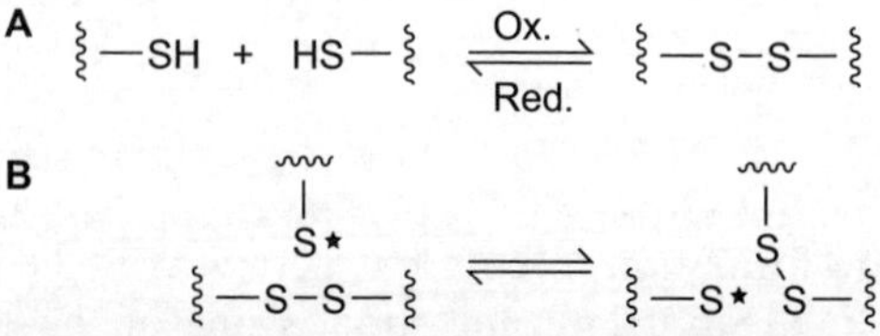

Figure 3.12   A) Oxidative disulfide formation from thiol functional group, and B) subsequent exchange reactions, where the star represents either a radical or anion.

molecules by utilizing a phosphine catalyst,[82,83] which is readily extended to disulfide-based CANs.

Photo-mediated connectivity rearrangement of a disulfide cross-linked PEG hydrogel was recently demonstrated by Fairbanks *et al.*[70] Here, the network was directly formed by the oxidation of thiols on a 4-armed PEG in water, using a hydrogel peroxide/sodium iodide solution. Photoinitiator was diffused into the hydrogel and the material was subsequently strained and photo-patterned (*i.e.* mechanophotopatterned). Utilizing this technique, the material's equilibrium geometry was manipulated using masked irradiation. Additionally, when two hydrogel specimens were pressed together and irradiated, the radical-induced rearrangement of disulfide bonds resulted in the fusion of the pieces into a single hydrogel piece, demonstrating the material's healing potential. It is noteworthy that the observed stress relaxation cannot simply be explained by photolytic cleavage of the disulfide bond; rather, the mechanism must follow an exchange reaction (Figure 3.12B), similar to addition–fragmentation of the RAFT-based materials (see Section 3.4.2).

3.4.4   The Diels-Alder Reaction

Of the various cycloaddition cross-linking reactions, the thermoreversible $[2+4]$ cycloaddition, or Diels-Alder (DA), reaction is among the more robust and easy to implement, and it even has the preeminent status of being a 'click' reaction.[17] The DA reaction is reasonably tolerant of a range of environmental conditions, such as water and oxygen, and the reactivity is readily tuned by the correct choice of conjugated diene and electronically activated double bond, or dienophile.[84–86]

The initial applications of the Diels-Alder reaction in CANs dates back to the 1960s, when Craven cross-linked a polymer having pendant furan functional groups with a multifunctional maleimide monomer (Figure 3.13A).[8] The thermoreversible cross-linked polymer is able to be post-formed upon heating the material to 120–140 °C. This thermoreversible cross-linking strategy has since been used by a number of researchers to produce a reversible macromer network architecture.[87–96] It should be noted that this material does not undergo reverse gelation at these temperatures, but is nevertheless remoldable (or healable) while roughly maintaining its structure. Typically, the reversible gelation temperature for these reversible macromers is impractically high, triggering irreversible side reactions that limit its healing ability.

Utilizing tetrafunctional furan and trifunctional maleimide monomers, Chen *et al.* demonstrated the healing capability of thermoreversible CANs using a reversible monomer network architecture strategy.[97] In this study, the material was cracked using fracture toughness testing and remended by heating the specimen for a prescribed length of time (*e.g.* 2 hours at 120 °C). During the thermal treatment, the material transformed from a glassy to rubbery polymer matrix (Tg~100 °C),[98] providing molecular mobility and enabling the DA adducts to rearrange. At 120 °C, the sample was able to maintain its shape and be remended, exhibiting a healing efficiency of up to 50%. At sufficiently

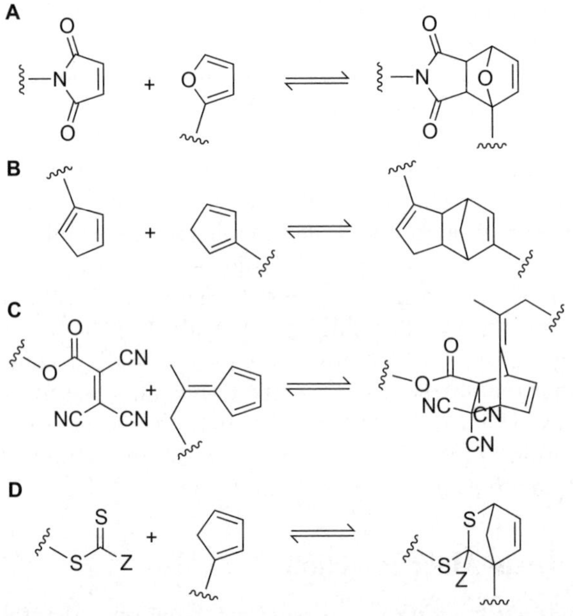

Figure 3.13   Select Diels–Alder cross-links: A) cyclopentadiene dimerization; B) maleimide–furan; C) tricyanoacrylate–fulvene; and D) RAFT agent–cyclopentadiene.

elevated temperatures the material would, in principle, undergo a gel-to-sol transition. Unfortunately, at these temperatures, such materials tend to undergo an irreversible side reaction that limits their healing capability over time. Nevertheless, the gel point conversion (*i.e.* the gel-to-sol transition) is readily tuned by changing the number of functional groups in accordance with the Flory-Stockmayer equation.[99]

Furan and maleimide exhibit excellent kinetic and thermodynamic properties for a range of healing materials applications; however, one of the strengths of the DA cross-linking approach is that the kinetic and thermodynamics are tuned by the choice of diene and dienophile. The ability of cyclopentadiene to act as both a diene and dienophile has been used to cross-link linear polymers *via* the dimerization of cyclopentadienyl pendant groups (Figure 3.13B).[100] Furthermore, if the cyclopentadiene is conjugated with an ester, the dicyclopentadiene dimer can undergo further reaction with cyclopentadiene to form a trimer, enabling a difunctional cyclopentadiene monomer to undergo cross-linking.[101] Lehn and co-workers have explored fulvene and cyanoethylene-carboxylate (Figure 3.13C) as a diene-dienophile pair that has fast reaction kinetics at room temperature.[102,103] Utilizing multifunctional tricyanoethylene-carboxylate and fulvene-based monomers, they demonstrated rapid (~ 10 seconds) self-healing by simply pressing two pieces of this material together.[103] Moreover, the material did not 'stick' to other materials, highlighting the selectivity of the reversible DA cross-linking reactions. Finally, Barner-Kowollick and co-workers have explored the hetero-DA reaction of a thio-thiocarbonyl 'RAFT' agent with cyclopentadiene (Figure 3.13D), where the kinetics were more rapid than those of the maleimide–furan cycloaddition.[104] The utilization of a difunctional RAFT agent enables the highly controlled synthesis of the polymer network backbone combined with facile reversible cross-linking using a multifunctional cyclopentadiene monomer.

3.4.5 Other Reversible Chemistries

In this subsection, several reversible chemistries that ultimately may have utility as a healing strategy in covalent networks are briefly mentioned. As mentioned previously, these reversible reactions have thus far not been considered for healing applications, or they exhibit limited reversibility. In the latter case, the reversible chemistries may be suitable for specialized healing applications, or may become practical with the utilization of a catalyst.

3.4.5.1 Ionenes

Ionenes are polymers that are formed from multifunctional halide and amine monomers (*i.e. via* the Menschutkin reaction), creating a quaternary ammonium cross-link (Figure 3.14).[105–107] The reaction necessarily creates a charged polymer, which is suitable for use in applications such as ion exchange membranes. The thermoreversible nature enables depolymerization at elevated temperatures.[108] Ruckenstein et al.[109] have utilized both a difunctional halide and an amine monomer to cross-link polymers having pendant amine and halide functional groups respectively. While the reversibility kinetics exhibited a significant dependence on the amine and halide used, most of the gelled materials were able to 'flow' at elevated temperatures (215 °C) despite the presence of a side reaction (*i.e.* Hoffman degradation) observed in some materials.[109]

3.4.5.2 Isocyanate–Imidazole Reactions

While a number of nucleophilic addition reactions are reversible, the nucleophile must also be a good leaving group for the reaction to be suitable for reversible cross-links. For instance, alcohols and amines react with isocyanates, but do not undergo the reverse reaction at practical temperatures. In contrast, imidazoles are capable of undergoing a nucleophilic addition reaction with an isocyanate, as well as the reversible reaction (Figure 3.15A). While the

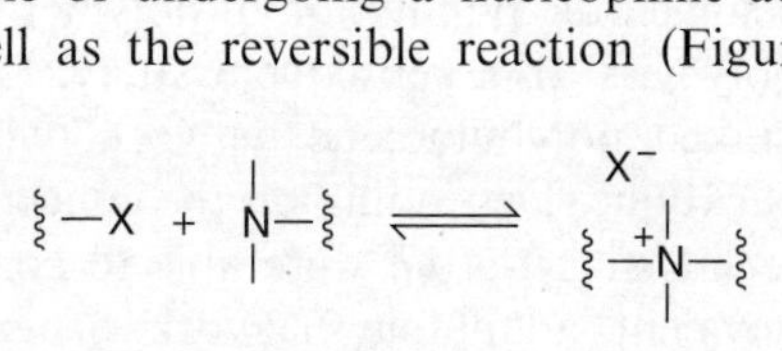

$$\xi\!-\!X \ + \ \underset{|}{\overset{|}{N}}\!-\!\xi \ \rightleftharpoons \ \xi\!-\!\overset{X^-}{\underset{|}{\overset{+}{N}}}\!-\!\xi$$

Figure 3.14   Ionene cross-link formation *via* the Menschutkin reaction between a halide (X) and a tertiary amine.

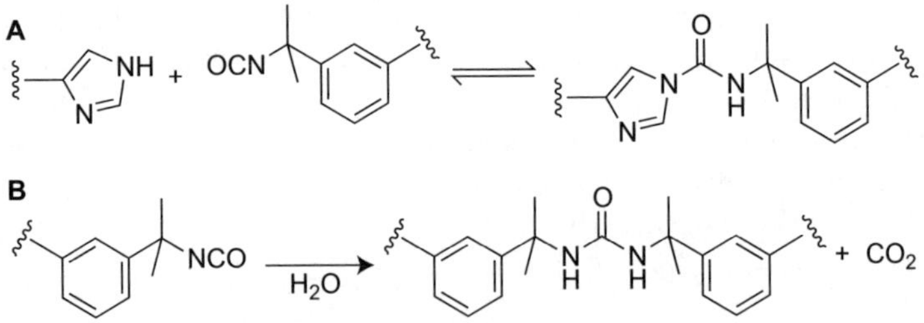

Figure 3.15 Thermoreversible cross-linking reaction between imidazole and isocyantate (A), and irreversible side reaction with water (B).

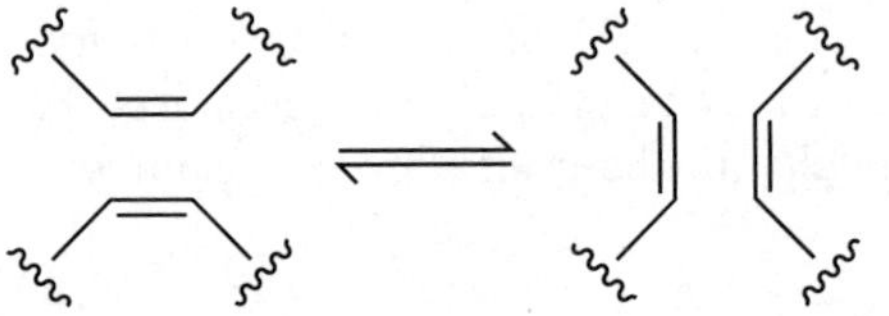

Figure 3.16 Olefin metathesis reaction.

isocyanate–imidazole reaction demonstrated good thermoreversible properties in a polymer network under inert conditions,[110] the isocyanate is, in general, intolerant of moisture. Unlike the imine-based networks, which reversibly depolymerize in the presence of water, an isocyanate will irreversibly form a urea cross-link (with itself) and release carbon dioxide upon exposure to water (Figure 3.15B).

3.4.5.3 *Olefin Metathesis*

Olefin metathesis has been extensively used in small molecule and polymer synthesis to create a number of new molecular architectures (Figure 3.16).[111,112] The introduction of specific metal-based catalysts (*e.g.* the ruthenium-based Grubbs' catalyst) enables the rapid interchange of olefin bonds that is an essential component of ring-opening metathesis polymerization (ROMP). In a recent demonstration of bond shuffling, Otsuka *et al.*[113] utilized two homopolymers, which contained olefins within the polymer backbone, and introduced Grubbs' catalyst into the polymer mixture. Over time, the catalyzed metathesis reaction changed the population of polymer chains into alternating units of both homopolymers. In a subsequent study, Lu *et al.*[114] utilized this concept in a cross-linked polybutadiene network, which contained olefins within the network backbone. Depending on the concentration of the second-generation Grubbs' catalyst, Lu *et al.* were able to control the rate of stress relaxation in this covalent adaptable network. This simple approach of inducing network rearrangement *via* metathesis is broadly applicable to the majority of polymer networks containing olefins.

3.4.5.4 Thermoreversible Nitroxide-mediated Reactions

The thermoreversible alkoxyamine bond has been extensively utilized in nitroxide-mediated polymerization (NMP) to fabricate a range of linear polymer architectures.[115–117] In this reaction, a stable nitroxide radical will react and essentially sequester a propagating radical, thereby limiting bimolecular radical termination. The reaction kinetics for the sequestering reaction are rapid, but the equilibrium is shifted towards the non-sequestered propagating radical at elevated temperatures.

Higaki *et al.*[118] recognized that the thermoreversible nitroxide mechanism could also be incorporated as a cross-link in a reversible network. In this work, they used pendent 2,2,6,6-tetramethylpiperidinyl-1-oxy (TEMPO) and styryl functional groups to cross-link poly(methyl methacrylate) linear chains (Figure 3.17). While a TEMPO-styryl small molecular species is initially produced upon cross-linking, in contrast to other reversible condensation reactions, its presence is not required for the polymer network to be thermoreversible. An additional advantage of this strategy is that the polymer network is readily modified by swelling in other polymerizable monomers.[119] Upon heating, the non-sequestered form of the propagating radical will be generated and incorporate the new monomer. This strategy has great potential for healing applications.

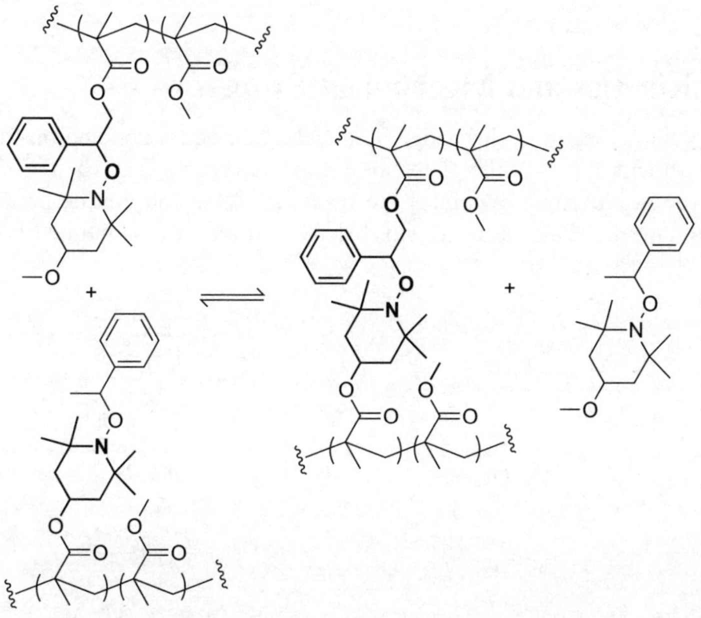

Figure 3.17 Thermorevesible cross-linking of the TEMPO-styryl functional groups.[118] Upon heating, the pendent styryl and nitroxide undergo a radical crossover reaction, producing a non-attached TEMPO-styryl compound.

3.4.5.5 Other Cycloadditions: The [2+2] and [4+4] Cycloadditions

Photoiniduced cycloaddition reactions are capable of photo-reversible cross-linking of polymeric materials. Irradiation of the material at one wavelength (typically near-UV) will induce the forward, photodimerization reaction, while a second, shorter wavelength induces the reverse, photoscission reaction (Figure 3.18). Most photocyclizations are of the [2+2] variety; products formed from this reation include: coumarins,[120–124] cinnamates,[125–129] and thymines.[130] Anthracenes,[127,131–133] however, are able to undergo the [4+4] photoinduced cycloaddition reaction. One of the unique aspects of these photoinduced reactions is that the reaction ceases once the irradiation is ceased; thus the state of cross-linking can be set by the wavelength and irradiation dose. In previous examples of photoinduced healing reactions, the irradiation resulted in the generation of radicals, which terminated on a relatively short timescale. This is not the case for these photodimerization reactions, which are a one-photon, one-reaction scheme; however, significant intensity is required to affect the network cross-linking. In contrast, addition–fragmentation and disulfide materials undergo a one-photon, reaction cascade scheme, requiring a substantially lower intensity to induce network rearrangement (see Sections 3.4.2 and 3.4.3). Additionally, typical photodimerization irradiation wavelengths are below 300 nm, which both limits the penetration depth of light and can initiate other side reactions in the material.

3.5 Kinetics and Mechanical Properties

In the previous sections, the design of a CAN focused on bond connectivity and rearrangement type. While these are of primary concern when designing a CAN, the microscopic dynamics are responsible for the unique properties of these materials. This section will briefly outline the dynamic mechanical

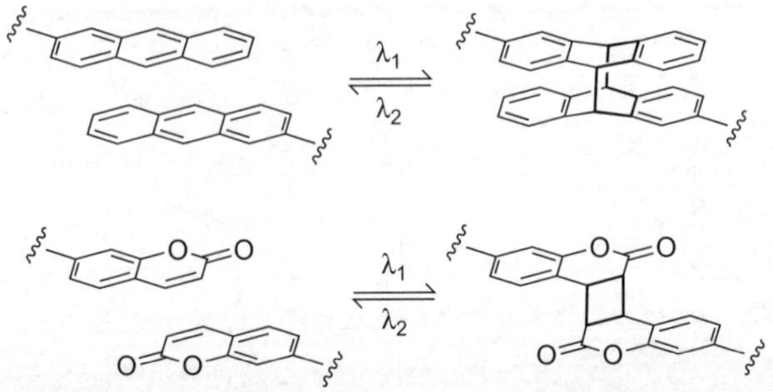

Figure 3.18 Photodimerization/photoscission of anthracene (top) and coumarin (bottom). The boldened bonds highlight the formed rings that define the [4+4] and [2+2] cycloadditions (top and bottom respectively). It should be noted that typical photodimerization reactions produce several isomers (*e.g.* head-to-head and head-to-tail, both *syn* and *anti*).

behavior associated with CANs and draw a loose connection to the underlying microscopic kinetics of bond rearrangement.

The most distinct characteristic of a reversible addition network is the apparent contradiction of a chemically cross-linked polymer that is capable of undergoing plastic deformation. That is, it possesses a gel-point, yet does not have a finite zero-frequency modulus. Owing to the dynamic nature of these materials, dynamic mechanical analysis (DMA) is often employed to probe their time-dependent mechanical response.

General to most covalent networks, CANs with reversible addition cross-links exhibit a rubbery 'plateau' modulus at intermediate frequencies, which increases sharply, corresponding to high frequency glassy dynamics (Figure 3.19).[134] In contrast to conventional covalent networks, these cross-linked materials exhibit characteristics of a viscoelastic liquid at low frequency (*i.e.* $G' \sim \omega^2$ and $G'' \sim \omega^1$, see Section 1.5.3).[99] The dynamic mechanical behavior in these CANs is, therefore, similar to that observed in linear polymer melts. Nevertheless, there are two critical differences between a polymer melt and a CAN: 1) the plateau modulus is related to the entanglements in a melt and the cross-link density in a CAN, and 2) the relaxation (*i.e.* low frequency behavior) is related to the curvilinear diffusion of a polymer chain (*i.e.* reptation) in a melt, and the reverse reaction kinetics of the cross-links in a CAN.[1,99]

Adzima *et al.*[99] critically examined both the chemical kinetic and dynamic mechanical properties of a thermoreversible Diels–Alder network. From this study, two key observations were made: 1) increasing temperature decreases the number of cross-links and thereby the modulus, and 2) the longest relaxation time of the network is related to the reverse reaction kinetics of the cross-link

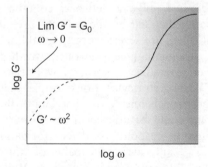

Figure 3.19 Idealized complex elastic modulus as a function of frequency for viscoelastic solid and liquid (solid and dashed line respectively). Covalent network mechanical behavior tends to follow that of a viscoelastic solid, where the zero-frequency modulus is directly related to the cross-link density through rubber elasticity theory. In contrast, a CAN elastic modulus tends to follow the behavior of the dashed line, which exhibits the relaxation of a viscoelastic liquid. Both viscoelastic solids and liquids exhibit a large increase in the elastic modulus at high frequency (shaded region), owing to glass dynamics.

(*i.e.* the Diels–Alder adduct half-life). These observations imply that one can predict the intermediate to low frequency portion of the DMA curve given the reaction kinetics of an addition-type, reversible monomer-network architecture.

Exchange reactions undergo cleavage and reformation in a manner that preserves the total number of cross-links; thus the position of the plateau modulus is less affected by temperature (*i.e.* other than that associated with the rubber elasticity theory). Nevertheless, the material should exhibit relaxation at long timescales (low frequencies). Indeed, Montarnal *et al.*[11] observed an almost temperature independent modulus (in fact there was an increase owing to the temperature dependence of rubber elasticity theory) and a significant decrease in relaxation time with increasing temperature.

The two examples of reversible addition and exchange reactions were for *monomer* network architectures. The relaxation behavior is expected to be complicated significantly by the convolution of the both reversible cross-links and relaxation *via* reptation in *macromer* network architectures. Nevertheless, such material systems are among the first CANs to have demonstrated tremendous remold and healing potential.

3.6   Conclusion and Outlook

The grand challenge in covalently cross-linked materials is to create smart materials that are capable of material healing analogous to that of biological systems. Similar to the way in which biological systems employ different strategies for healing bone fractures *versus* bruised or punctured soft tissue, different material designs must be considered for each healing application.[135] The dynamic properties of polymer networks containing reversible covalent cross-links are primarily dictated by the network architecture and reversible cross-link type. While there are several reversible addition- and exchange-type covalent chemistries, as highlighted previously, further implementations of DCC or other novel reversible chemistries will lead to new innovations in covalent network healing strategies. For example, there is a great potential for the role of mechanochemistry to accelerate the adaptability of a polymer network. This could be utilized in cases where the reaction kinetics are not rapid enough to respond to defect formation and propagation. Mechanochemistry[135–137] may even be capable of triggering other reactions that will locally heal a material where it is weakest. With these new materials, there is a significant need for better characterization of the reaction kinetics, in concert with the dynamic mechanical properties, to gain new insights in CAN design. Nevertheless, the foundational concepts and tools are available for the *a priori* design of a CAN for a specific healing application. Utilizing reversible cross-links within a polymer network framework has defined a new class of smart materials that will ultimately enhance the service lifetime of polymeric materials in a wide range of advanced materials applications.

References

1. C. J. Kloxin, T. F. Scott, B. J. Adzima and C. N. Bowman, *Macromolecules*, 2010, **43**, 2643.
2. L. P. Engle and K. B. Wagener, *J. Macromol. Sci. Rev. Macomol. Chem. Phys.*, 1993, **C33**, 239.
3. J. M. Lehn, *Chem. Eur. J.*, 1999, **5**, 2455.
4. M. E. Belowich and J. F. Stoddart, *Chem. Soc. Rev.*, 2012, **41**, 2003.
5. S. J. Rowan, S. J. Cantrill, G. R. L. Cousins, J. K. M. Sanders and J. F. Stoddart, *Angew. Chem. Int. Ed.*, 2002, **41**, 898.
6. P. J. Flory, *Chem. Rev.*, 1944, **35**, 51.
7. M. S. Green and A. V. Tobolsky, *J. Chem. Phys.*, 1946, **14**, 80.
8. J. M. Craven, Dupont, US Patent #3,435,003, 1969.
9. J. M. Lehn, *Prog. Polym. Sci.*, 2005, **30**, 814.
10. M. Capelot, D. Montarnal, F. Tournilhac and L. Leibler, *J. Am. Chem. Soc.*, 2012, **134**, 7664.
11. D. Montarnal, M. Capelot, F. Tournilhac and L. Leibler, *Science*, 2011, **334**, 965.
12. C. A. Angell, *Science*, 1995, **267**, 1924.
13. A. B. S. Bakhtiari, D. Hsiao, G. X. Jin, B. D. Gates and N. R. Branda, *Angew. Chem. Int. Ed.*, 2009, **48**, 4166.
14. S. Yamashita, H. Fukushima, Y. Niidome, T. Mori, Y. Katayama and T. Niidome, *Langmuir*, 2011, **27**, 14621.
15. B. J. Adzima, C. J. Kloxin and C. N. Bowman, *Adv. Mater.*, 2010, **22**, 2784.
16. R. J. Wojtecki, M. A. Meador and S. J. Rowan, *Nat. Mater.*, 2011, **10**, 14.
17. H. C. Kolb, M. G. Finn and K. B. Sharpless, *Angew. Chem. Int. Ed.*, 2001, **40**, 2004.
18. A. Ausin, I. Eguiazabal, M. E. Munoz, J. J. Pena and A. Santamaria, *Polym. Eng. Sci.*, 1987, **27**, 529.
19. P. J. Flory, *J. Am. Chem. Soc.*, 1940, **62**, 2255.
20. P. J. Flory, *J. Am. Chem. Soc.*, 1940, **62**, 2261.
21. R. S. Porter and L. H. Wang, *Polymer*, 1992, **33**, 2019.
22. A. R. Gilbert and S. W. Kantor, *J. Polym. Sci.*, 1959, **40**, 35.
23. S. W. Kantor, W. T. Grubb and R. C. Osthoff, *J. Am. Chem. Soc.*, 1954, **76**, 5190.
24. P. W. Zheng and T. J. McCarthy, *J. Am. Chem. Soc.*, 2012, **134**, 2024.
25. R. C. Osthoff, A. M. Bueche and W. T. Grubb, *J. Am. Chem. Soc.*, 1954, **76**, 4659.
26. R. W. Layer, *Chem. Rev.*, 1963, **63**, 489.
27. C. D. Meyer, C. S. Joiner and J. F. Stoddart, *Chem. Soc. Rev.*, 2007, **36**, 1705.
28. N. C. Duncan, B. P. Hay, E. W. Hagaman and R. Custelcean, *Tetrahedron*, 2012, **68**, 53.
29. P. Pandey, A. P. Katsoulidis, I. Eryazici, Y. Y. Wu, M. G. Kanatzidis and S. T. Nguyen, *Chem. Mater.*, 2010, **22**, 4974.

30. (a) F. J. Uribe-Romo, J. R. Hunt, H. Furukawa, C. Klock, M. O'Keeffe and O. M. Yaghi, *J. Am. Chem. Soc.*, 2009, **131**, 4570; (b) T. Hasell, M. Schmidtnamm, C. C. A. Stone, M. W. Smith and A. I. Cooper, *Chem. Commun.*, 2012, **48**, 4689.

31. P. Kovaricek and J. M. Lehn, *J. Am. Chem. Soc.*, 2012, **134**, 9446.

32. K. Osowska and O. S. Miljanic, *J. Am. Chem. Soc.*, 2011, **133**, 724.

33. D. Colombani and P. Chaumont, *Prog. Polym. Sci.*, 1996, **21**, 439.

34. T. F. Scott, A. D. Schneider, W. D. Cook and C. N. Bowman, *Science*, 2005, **308**, 1615.

35. D. N. Hall, *J. Org. Chem.*, 1967, **32**, 2082.

36. D. N. Hall, A. A. Oswald and K. Griesbau, *J. Org. Chem.*, 1965, **30**, 3829.

37. T. Migita, M. Kosugi, K. Takayama and Y. Nakagawa, *Tetrahedron*, 1973, **29**, 51.

38. D. H. R. Barton and D. Crich, *J. Chem. Soc., Perkin Trans.*, 1986, **1**, 1613.

39. G. F. Meijs, E. Rizzardo and S. H. Thang, *Macromolecules*, 1988, **21**, 3122.

40. G. F. Meijs and E. Rizzardo, *Makromol. Chem., Rapid Commun.*, 1988, **9**, 547.

41. P. Cacioli, D. G. Hawthorne, R. L. Laslett, E. Rizzardo and D. H. Solomon, *J. Macromol. Sci. Chem.*, 1986, **A23**, 839.

42. J. Chiefari, Y. K. Chong, F. Ercole, J. Krstina, J. Jeffery, T. P. T. Le, R. T. A. Mayadunne, G. F. Meijs, C. L. Moad, G. Moad, E. Rizzardo and S. H. Thang, *Macromolecules*, 1998, **31**, 5559.

43. R. A. Evans and E. Rizzardo, *J. Polym. Sci. Pol. Chem.*, 2001, **39**, 202.

44. R. A. Evans and E. Rizzardo, *Macromolecules*, 1996, **29**, 6983.

45. R. A. Evans, G. Moad, E. Rizzardo and S. H. Thang, *Macromolecules*, 1994, **27**, 7935.

46. C. J. Kloxin, T. F. Scott and C. N. Bowman, *Macromolecules*, 2009, **42**, 2551.

47. H. Y. Park, C. J. Kloxin, M. F. Fordney and C. N. Bowman, *Macromolecules*, 2012, **45**, 5647.

48. H. Y. Park, C. J. Kloxin, A. S. Abuelyaman, J. D. Oxman and C. N. Bowman, *Macromolecules*, 2012, **45**, 5640.

49. H. Y. Park, C. J. Kloxin, T. F. Scott and C. N. Bowman, *Macromolecules*, 2010, **43**, 10188.

50. H. Y. Park, C. J. Kloxin, T. F. Scott and C. N. Bowman, *Dent. Mater.*, 2010, **26**, 1010.

51. C. J. Kloxin, T. F. Scott, H. Y. Park and C. N. Bowman, *Adv. Mater.*, 2011, **23**, 1977.

52. Y. Amamoto, J. Kamada, H. Otsuka, A. Takahara and K. Matyjaszewski, *Angew. Chem. Int. Ed.*, 2011, **50**, 1660.

53. R. Nicolay, J. Kamada, A. Van Wassen and K. Matyjaszewski, *Macromolecules*, 2010, **43**, 4355.

54. R. T. A. Mayadunne, J. Jeffery, G. Moad and E. Rizzardo, *Macromolecules*, 2003, **36**, 1505.

55. G. Moad, J. Chiefari, Y. K. Chong, J. Krstina, R. T. A. Mayadunne, A. Postma, E. Rizzardo and S. H. Thang, *Polym. Int.*, 2000, **49**, 993.
56. C. Boyer, V. Bulmus and T. P. Davis, *Macromol. Rapid Commun.*, 2009, **30**, 493.
57. S. R. Gondi, A. P. Vogt and B. S. Sumerlin, *Macromolecules*, 2007, **40**, 474.
58. M. Hales, C. Barner-Kowollik, T. P. Davis and M. H. Stenzel, *Langmuir*, 2004, **20**, 10809.
59. Y. Amamoto, H. Otsuka, A. Takahara and K. Matyjaszewski, *ACS Macro Lett.*, 2012, **1**, 478.
60. Y. Z. You, C. Y. Hong, R. K. Bai, C. Y. Pan and J. Wang, *Macromol. Chem. Phys.*, 2002, **203**, 477.
61. Y. Amamoto, H. Otsuka, A. Takahara and K. Matyjaszewski, *Adv. Mater.*, 2012.
62. G. Moad, E. Rizzardo and S. H. Thang, *Aust. J. Chem.*, 2009, **62**, 1402.
63. G. Moad, E. Rizzardo and S. H. Thang, *Aust. J. Chem.*, 2006, **59**, 669.
64. G. Moad, E. Rizzardo and S. H. Thang, *Aust. J. Chem.*, 2005, **58**, 379.
65. G. Moad, E. Rizzardo and S. H. Thang, *Polymer*, 2008, **49**, 1079.
66. A. V. Tobolsky, W. J. Macknight and M. Takahashi, *J. Phys. Chem.*, 1964, **68**, 787.
67. Y. Takahash. and A. V. Tobolsky, *Polym. J.*, 1971, **2**, 457.
68. J. A. Yoon, J. Kamada, K. Koynov, J. Mohin, R. Nicolay, Y. Z. Zhang, A. C. Balazs, T. Kowalewski and K. Matyjaszewski, *Macromolecules*, 2012, **45**, 142.
69. G. H. Deng, F. Y. Li, H. X. Yu, F. Y. Liu, C. Y. Liu, W. X. Sun, H. F. Jiang and Y. M. Chen, *ACS Macro Lett.*, 2012, **1**, 275.
70. B. D. Fairbanks, S. P. Singh, C. N. Bowman and K. S. Anseth, *Macromolecules*, 2011, **44**, 2444.
71. J. Canadell, H. Goossens and B. Klumperman, *Macromolecules*, 2011, **44**, 2536.
72. S. Cerritelli, D. Velluto and J. A. Hubbell, *Biomacromolecules*, 2007, **8**, 1966.
73. M. Pooga, U. Soomets, M. Hallbrink, A. Valkna, K. Saar, K. Rezaei, U. Kahl, J. X. Hao, X. J. Xu, Z. Wiesenfeld-Hallin, T. Hokfelt, T. Bartfai and U. Langel, *Nat. Biotechnol.*, 1998, **16**, 857.
74. G. Saito, J. A. Swanson and K. D. Lee, *Adv. Drug Deliv. Rev.*, 2003, **55**, 199.
75. S. Takae, K. Miyata, M. Oba, T. Ishii, N. Nishiyama, K. Itaka, Y. Yamasaki, H. Koyama and K. Kataoka, *J. Am. Chem. Soc.*, 2008, **130**, 6001.
76. O. Ramstrom and J. M. Lehn, *ChemBioChem*, 2000, **1**, 41.
77. G. Dalman, J. McDermed and G. Gorin, *J. Org. Chem.*, 1964, **29**, 1480.
78. A. Fava, A. Iliceto and E. Camera, *J. Am. Chem. Soc.*, 1957, **79**, 833.
79. J. Houk and G. M. Whitesides, *J. Am. Chem. Soc.*, 1987, **109**, 6825.
80. N. V. Tsarevsky and K. Matyjaszewski, *Macromolecules*, 2002, **35**, 9009.
81. Y. Chujo, K. Sada, A. Naka, R. Nomura and T. Saegusa, *Macromolecules*, 1993, **26**, 883.

82. R. Caraballo, M. Rahm, P. Vongvilai, T. Brinck and O. Ramstrom, *Chem. Commun.*, 2008, 6603.

83. R. Caraballo, M. Sakulsombat and O. Ramstrom, *Chem. Commun.*, 2010, **46**, 8469.

84. R. C. Boutelle and B. H. Northrop, *J. Org. Chem.*, 2011, **76**, 7994.

85. K. C. Koehler, A. Durackova, C. J. Kloxin and C. N. Bowman, *AIChE J.*, 2012, **58**, 3545.

86. F. Fringuelli and A. Taticchi, *The Diels-Alder Reaction: Selected Practical Methods*, John Wiley & Sons, Inc., New York, 2002.

87. H. Laita, S. Boufi and A. Gandini, *Eur. Polym. J.*, 1997, **33**, 1203.

88. C. Gousse, A. Gandini and P. Hodge, *Macromolecules*, 1998, **31**, 314.

89. R. Gheneim, C. Perez-Berumen and A. Gandini, *Macromolecules*, 2002, **35**, 7246.

90. E. Goiti, F. Heatley, M. B. Huglin and J. M. Rego, *Eur. Polym. J.*, 2004, **40**, 1451.

91. E. Goiti, M. B. Huglin and J. M. Rego, *Eur. Polym. J.*, 2004, **40**, 219.

92. Y. L. Liu and Y. W. Chen, *Macromol. Chem. Phys.*, 2007, **208**, 224.

93. Y. L. Liu and C. Y. Hsieh, *J. Polym. Sci. Pol. Chem.*, 2006, **44**, 905.

94. A. A. Kavitha and N. K. Singha, *Macromol. Chem. Phys.*, 2007, **208**, 2569.

95. Y. Zhang, A. A. Broekhuis and F. Picchioni, *Macromolecules*, 2009, **42**, 1906.

96. Y. Imai, H. Itoh, K. Naka and Y. Chujo, *Macromolecules*, 2000, **33**, 4343.

97. X. X. Chen, M. A. Dam, K. Ono, A. Mal, H. B. Shen, S. R. Nutt, K. Sheran and F. Wudl, *Science*, 2002, **295**, 1698.

98. X. X. Chen, F. Wudl, A. K. Mal, H. B. Shen and S. R. Nutt, *Macromolecules*, 2003, **36**, 1802.

99. B. J. Adzima, H. A. Aguirre, C. J. Kloxin, T. F. Scott and C. N. Bowman, *Macromolecules*, 2008, **41**, 9112.

100. J. P. Kennedy and K. F. Castner, *J. Polym. Sci. Pol. Chem.*, 1979, **17**, 2055.

101. E. B. Murphy, E. Bolanos, C. Schaffner-Hamann, F. Wudl, S. R. Nutt and M. L. Auad, *Macromolecules*, 2008, **41**, 5203.

102. P. J. Boul, P. Reutenauer and J. M. Lehn, *Org. Lett.*, 2005, **7**, 15.

103. P. Reutenauer, P. J. Boul and J. M. Lehn, *Eur. J. Org. Chem.*, 2009, 1691.

104. A. J. Inglis, S. Sinnwell, M. H. Stenzel and C. Barner-Kowollik, *Angew. Chem. Int. Ed.*, 2009, **48**, 2411.

105. A. Rembaum, W. Baumgartner and A. Eisenberg, *J. Polym. Sci. B Polym. Lett.*, 1968, **6**, 159.

106. C. F. Gibbs, E. R. Littmann and C. S. Marvel, *J. Am. Chem. Soc.*, 1933, **55**, 753.

107. S. R. Williams and T. E. Long, *Prog. Polym. Sci.*, 2009, **34**, 762.

108. C. M. Leir and J. E. Stark, *J. Appl. Polym. Sci.*, 1989, **38**, 1535.

109. E. Ruckenstein and X. N. Chen, *Macromolecules*, 2000, **33**, 8992.

110. J. Y. Chang, S. K. Do and M. J. Han, *Polymer*, 2001, **42**, 7589.

111. R. H. Grubbs, *Tetrahedron*, 2004, **60**, 7117.

112. R. H. Grubbs and S. Chang, *Tetrahedron*, 1998, **54**, 4413.

113. H. Otsuka, T. Muta, M. Sakada, T. Maeda and A. Takahara, *Chem. Commun.*, 2009, 1073.
114. Y. X. Lu, F. Tournilhac, L. Leibler and Z. B. Guan, *J. Am. Chem. Soc.*, 2012, **134**, 8424.
115. W. A. Braunecker and K. Matyjaszewski, *Prog. Polym. Sci.*, 2007, **32**, 93.
116. C. J. Hawker, G. G. Barclay, A. Orellana, J. Dao and W. Devonport, *Macromolecules*, 1996, **29**, 5245.
117. C. J. Hawker, A. W. Bosman and E. Harth, *Chem. Rev.*, 2001, **101**, 3661.
118. Y. Higaki, H. Otsuka and A. Takahara, *Macromolecules*, 2006, **39**, 2121.
119. Y. Amamoto, M. Kikuchi, H. Masunaga, S. Sasaki, H. Otsuka and A. Takahara, *Macromolecules*, 2009, **42**, 8733.
120. S. R. Trenor, A. R. Shultz, B. J. Love and T. E. Long, *Chem. Rev.*, 2004, **104**, 3059.
121. M. Nagata and Y. Yamamoto, *React. Funct. Polym.*, 2008, **68**, 915.
122. M. Nagata and Y. Yamamoto, *J. Polym. Sci. Pol. Chem.*, 2009, **47**, 2422.
123. D. L. Zhao, B. Y. Ren, S. S. Liu, X. X. Liu and Z. Tong, *Chem. Commun.*, 2006, 779.
124. B. Y. Ren, D. L. Zhao, S. S. Liu, X. X. Liu and Z. Tong, *Macromolecules*, 2007, **40**, 4501.
125. A. Lendlein, H. Y. Jiang, O. Junger and R. Langer, *Nature*, 2005, **434**, 879.
126. H. Y. Jiang, S. Kelch and A. Lendlein, *Adv. Mater.*, 2006, **18**, 1471.
127. Y. J. Zheng, F. M. Andreopoulos, M. Micic, Q. Huo, S. M. Pham and R. M. Leblanc, *Adv. Funct. Mater.*, 2001, **11**, 37.
128. M. Micic, Y. J. Zheng, V. Moy, X. H. Zhang, F. M. Andreopoulos and R. M. Leblanc, *Colloid Surf., B*, 2003, **27**, 147.
129. K. M. Gattas-Asfura, E. Weisman, F. M. Andreopoulos, M. Micic, B. Muller, S. Sirpal, S. M. Pham and R. M. Leblanc, *Biomacromolecules*, 2005, **6**, 1503.
130. Y. Imai, T. Ogoshi, K. Naka and Y. Chujo, *Polym. Bull.*, 2000, **45**, 9.
131. Y. J. Zheng, M. Mieie, S. V. Mello, M. Mabrouki, F. M. Andreopoulos, V. Konka, S. M. Pham and R. M. Leblanc, *Macromolecules*, 2002, **35**, 5228.
132. L. A. Connal, R. Vestberg, C. J. Hawker and G. G. Qiao, *Adv. Funct. Mater.*, 2008, **18**, 3315.
133. J. Matsui, Y. Ochi and K. Tamaki, *Chem. Lett.*, 2006, **35**, 80.
134. M. Rubinstein and R. H. Colby, *Polymer Physics*, Oxford University Press, Inc., New York, 2003.
135. A. Piermattei, S. Karthikeyan and R. P. Sijbesma, *Nat. Chem.*, 2009, **1**, 133.
136. M. M. Caruso, D. A. Davis, Q. Shen, S. A. Odom, N. R. Sottos, S. R. White and J. S. Moore, *Chem. Rev.*, 2009, **109**, 5755.
137. D. A. Davis, A. Hamilton, J. L. Yang, L. D. Cremar, D. Van Gough, S. L. Potisek, M. T. Ong, P. V. Braun, T. J. Martinez, S. R. White, J. S. Moore and N. R. Sottos, *Nature*, 2009, **459**, 68.

CHAPTER 4

Healable Supramolecular Polymeric Materials

BARNABY W. GREENLAND,*[a] GINA L. FIORE,[b]
STUART J. ROWAN[c] AND CHRISTOPH WEDER[b]

[a] Department of Chemistry, The University of Reading, Whiteknights, Reading, RG6 6AD UK; [b] Adolphe Merkle Institute, University of Fribourg, CH-1700 Fribourg, Switzerland; [c] Department of Macromolecular Science and Engineering, Case Western Reserve University, 2100 Adelbert Road, Cleveland, Ohio 44106-7202 USA
*Email: b.w.greenland@reading.ac.uk

4.1 Introduction

Over the course of the last century, synthetic polymers have revolutionized our daily lives. Their broad range of useful properties, which can be designed virtually at will, combined with their ease of processing and low cost, has led to their ubiquitous use in a range of applications, from plastic bags to electronics, automotive parts, food packaging and many others. Upon long-term exposure to environmental conditions, such as chemical, optical, mechanical, thermal and other stresses polymeric materials usually degrade and eventually fail. Mechanical failure of polymeric materials is often the result of crack formation and propagation until the material eventually completely fails.[1] To address this problem, researchers have recently developed several intriguing approaches to creating self-healing or healable polymers, which have the ability to repair themselves autonomously (self-healing materials), or can be healed upon exposure to an external stimulus such as heat, light, pressure or mechanical

RSC Polymer Chemistry Series No. 5
Healable Polymer Systems
Edited by Wayne Hayes and Barnaby W Greenland
© The Royal Society of Chemistry 2013
Published by the Royal Society of Chemistry, www.rsc.org

stress (healable materials).[2–9] It is evident that polymers which can be repaired after sustaining damage are poised to offer extended functionality and lifetime in a wide range of applications; this area has therefore begun to attract considerable attention in academic and industrial research laboratories around the world.

One simple way to repair cracks or scratches in thermoplastic polymers is to expose them to solvents or heat, which permits surface rearrangement, wetting, diffusion, and re-entanglement of the polymer chains.[10] The rates for the latter two processes are inversely proportional to the molecular weight, which makes the healing process generally slow and not practical on a reasonable timescale.[11] Additionally, this approach often requires that large sections of the material are treated in order to mend the damaged areas.

Stress-induced healing (see Chapter 2), where mechanical stresses not only lead to mechanical damage, but at the same time trigger a mechanism to heal the defects introduced, represents the first sophisticated approach to impart healability to polymeric materials. The first synthesis of such materials, which are detailed elsewhere in this book, involved the introduction of monomer-filled capsules,[12] along with an appropriate catalyst, into the polymer matrix. Upon mechanical damage, the capsules rupture and release the monomer, which fills the defects and serves as the healing agent that cures when it comes into contact with the catalyst.[13] As this process is initiated by crack formation/propagation, it is essentially autonomous and can be considered as an example of true self-healing. Although the healing occurs immediately, this process is generally not repeatable, since the healing agent is depleted. Therefore multiple healing events in the same damaged area is usually difficult with this approach, although the development of vascular network systems somewhat alleviate this issue.[14,15]

Stimuli-responsive polymers, which are assembled on the basis of dynamic, reversible bonds and can be disassembled upon exposure to an external stimulus, have emerged as an important family of materials in which repeated healing is possible.[3,7,16,17] This chapter serves to summarize the development of supramolecular polymer systems that are either self-healing or healable, whereby damage recovery is usually achieved by application of a suitable external stimuli. By way of introduction, the following section details some of the early key studies concerning supramolecular polymers and will aid the understanding of the optimization of this fascinating class of healable materials over recent years.

4.1.1 Supramolecular Polymer Chemistry: A Concise History

Inspired by the exquisite complex assemblies found in nature, much of the early work in supramolecular chemistry was focused on the formation of hydrogen bonded arrays. Consequently, supramolecular polymers consisting of oligomeric units with highly optimized hydrogen bonding end-groups have become the most intensely studied materials.[18]

During the earliest days of this field of research, work focused on understanding the properties of hydrogen bonded polymers in dilute solution

($\gg 10\%$ w/v). In this concentration regime, the degree of polymerization (DP) for the resulting supramolecular polymer was demonstrated[19] to be related to the concentration, [M], of the discrete supramolecular units and their respective binding constants, K_a, by Equation (1).

$$DP \approx (K_a[M])^{0.5} \tag{1}$$

From Equation (1), it follows that at millimolar concentrations, an appreciable DP for the supramolecular polymer is only achieved when the hydrogen bonding end-groups have extremely high binding constants ($<10^6 \, M^{-1}$). These values can only be achieved by producing highly designed systems containing multiple complementary hydrogen bonding interactions. Key early work in this field was carried out by the groups of Meijer and Sijbesma,[20] Zimmermann[21] and Lehn,[22] and resulted in the development of synthetically accessible supramolecular motifs that can be appended to a range of oligomeric building blocks (Figure 4.1, 1 to 3 respectively).

Hydrogen bonding has been the most studied supramolecular interaction from both a fundamental and an application driven viewpoint. However, a multitude of sophisticated non-covalent interactions have been investigated to excellent effect over the past 30 years. Two of the most distinctive are π–π stacking[23] and metal–ligand interaction,[24] both of which have found application in the field of healable supramolecular polymer chemistry.

π–π Stacking[23] can be observed between aromatic residues and is even observed in crystals of benzene in the solid state. However, the strength and directionality of the interactions can be greatly enhanced by tailoring the electronic properties of two different aromatics components to include a *donor* component that is π-electron rich, and an *acceptor* component that is π-electron deficient. One of the most well studied motifs that harnesses these interactions is the complex between π-electron deficient *N,N'*-dimethyl-4,4'-bipyridinium dicationic salts and substituted aromatics such as 1,4-dihydroxybenzene and 1,5-dihydroxynaphthalene, which are π-electron rich (Figure 4.2).[25,26] These donor–acceptor interactions have been harnessed to self-assemble complex

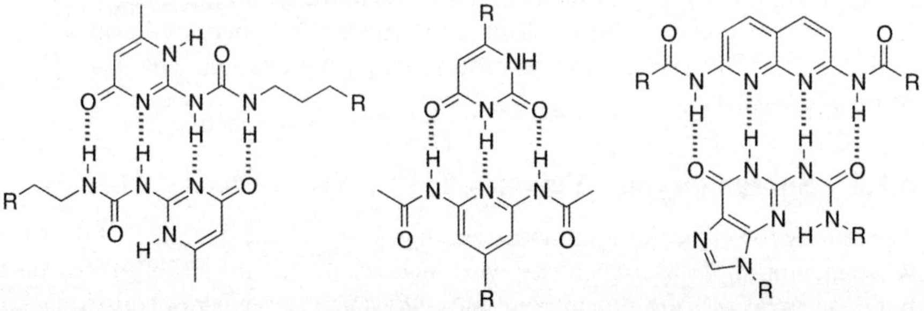

Figure 4.1 Structures of the supramolecular motifs developed by the groups of Meijer and Sijbesma, Zimmermann and Lehn (**1** to **3** respectively). Hydrogen bonding interactions are shown as blue dashes.

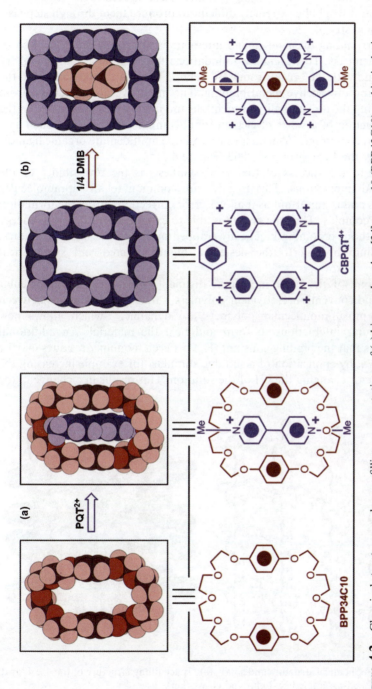

Figure 4.2 Chemical structures and space filling representations of (a) bisparaphenylene[34]crown 10 (BPP43C10) and its inclusion complex with paraquat (PQT2 +); (b) cyclobispara-*p*-phenylene (CBPGT) and its inclusion complex with 1,4-dimethoxybenzene (1/4 DMB).
Reproduced with permission from reference [26].

molecular machines and logic gates,[27] as well as 'daisy chains'[28,29] and hyper-branched polymers.[30] The resulting products have structures so ordered and complex that it would be extremely difficult to produce them through stepwise, covalent chemistry.

An understanding of metal–ligand interactions underpins the field of coordination chemistry. However, the foundations for supramolecular chemistry were also laid with the breakthrough by Lehn,[31] Pederson[32] and Cram[33] in the understanding of selective metal binding. The predictive power derived from understanding the directionality of metal–ligand interactions culminated in the rapid production of a series of elegant, multicomponent structures, such as a 'grid' system containing 16 metal ions and 8 large multidentate organic ligands, produced by the Lehn group in 2003 (Figure 4.3).[34]

Essential to the success of the 'grid' synthesis is the reversibility of the metal–ligand interactions. This is a feature common to all supramolecular interactions and is important as it allows error correction during the formation of the final complex. Thus the system is able to explore a series of local energy minima, without becoming kinetically trapped in any of the undesired states, whilst cycling through to the designed, highly ordered and symmetrical structure.

Indeed, one of the most important distinctions between supramolecular polymers and conventional, covalent polymers is that the associations between monomers in supramolecular polymers are dynamic,[35] which allows new supramolecular interactions to form and heal the material. An additional advantage is that the binding constant (K_a) between monomer residues may be altered *in situ* by application of a suitable stimulus, for example increasing the temperature or changing the pH of the solution. This results in a change in the

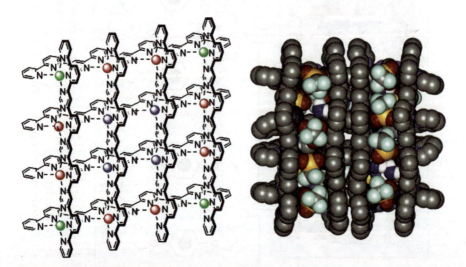

Figure 4.3   Chemical structure and solid-state space filling structure of the 4 × 4 'grid' self-assembled by Lehn and co-workers.
Adapted with permission from reference [31].

degree of polymerization of the supramolecular polymer [Equation (1)], and therefore a change in the physical properties of the solution, typically observed as a change in viscosity. Since covalent bonds have not been made or broken, removal of the stimulus restores the original properties of the solution. It is the reversible, tunable nature of supramolecular systems that makes them ideally suited to the production of healable materials.

The following section describes how this concept has been adapted to produce healable, supramolecular hydrogen bonded materials.

4.2 Healable Supramolecular Polymers Utilizing Hydrogen Bonding

Following the pioneering studies by the Meijer group concerning the synthesis and optimization of the ureidopyrimidinone (1) supramolecular end-group (Figure 4.1) which has become the pre-eminent motif for a broad range of supramolecular applications. Indeed, its ease of introduction in end-capping a multitude of oligomeric base polymers has resulted in the production of thermoreversible supramolecular polymeric materials on an industrial scale by Suprapolix BV®.[36] By heating such end-capped materials to 140 °C Suprapolix BV® have demonstrated crack healing visually in these films (Figure 4.4).

The relationship between DP and K_a *in solution* [Equation 1] was the primary driving force behind the design and synthesis of the highly-functionalized supramolecular motifs shown in Figure 4.1. However, most industrially important polymeric products, from plastic bottles to mobile telephone (cell phone) screens, are clearly intended to be used in the solid state, where the connection between DP and K_a is less well understood. Indeed, other factors, including microphase separation and crystallinity, have been shown to be major determinants of the physical characteristics of the product.

Section 4.2.2 details the realization of supramolecular polymers that exploit simpler hydrogen bonding residues with weaker and less well defined interactions to produce healable materials.

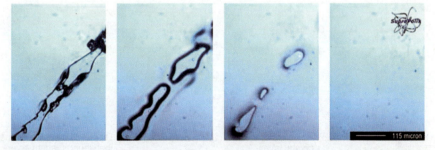

Figure 4.4 Crack healing in the surface of the supramolecular material produced by Suprapolix BV®.
Reproduced with permission from Suprapolix BV®.

4.2.1 Supramolecular Polymers Containing Hydrogen Bonding Motifs with Low Association Constants

In a series of reports published throughout the late 1980s and early 1990s, the Stadler group described the effect of introducing weakly hydrogen bonding residues, polybutadiene (PB), into a non-polar polymer backbone.[37-43] One of their most highly studied systems contained a benzoic acid modified urazole hydrogen bonding motif that contained a polymerizable vinyl group (**4**, Figure 4.5). Analysis of the solid-state structure of **4** revealed a two dimensional hydrogen bonding array consisting of alternative carboxylic acid dimers and urazoleresides. Dipole–dipole interactions between the carbonyl residues of the urazoles linked the layers into a three-dimensional network (Figure 4.5).

Copolymers of urazole monomer **4** (Figure 4.5) with butadiene were found to be thermoplastic elastomers. They behaved in a thermorheologically complex manner, suggesting a phase separated morphology, even at extremely low levels of urazole loading (>4 mol%).[39] Stadler proposed that the hydrogen bonding units were aggregating into discrete "hard" domains, embedded within larger, soft domains of polybutadiene, which accounted for 96% of the repeat units within the backbone of the polymers (Figure 4.6). Most importantly, it was demonstrated that the viscous-elastic transition for these copolymers (determined rheologically) was essentially coincident with the melting point for the hard segments (as determined by differential scanning calorimetry).[39] Thus the thermoresponse of the bulk polymer was entirely governed by the very weak (but concentrated) interactions of the hydrogen bonding units.

In 1992, Lillya and co-workers modified commercially available, low molecular weight poly(tetrahydrofuran) (pTHF, $M_n \approx 2000 \, \text{g mol}^{-1}$, $n \approx 28$) by the addition of terephthalic acid residues to generate two novel polymers that contained either a free acid group (**5**) or a benzyl protected acid residue (**6**) (Figure 4.7).[44] Both polymers were found to be highly thermoresponsive but exhibited dramatically different changes in their dynamic moduli (G', see Section 1.5.3) as a consequence of increasing temperature. Despite the close structural relationship between the polymers, over the temperature range

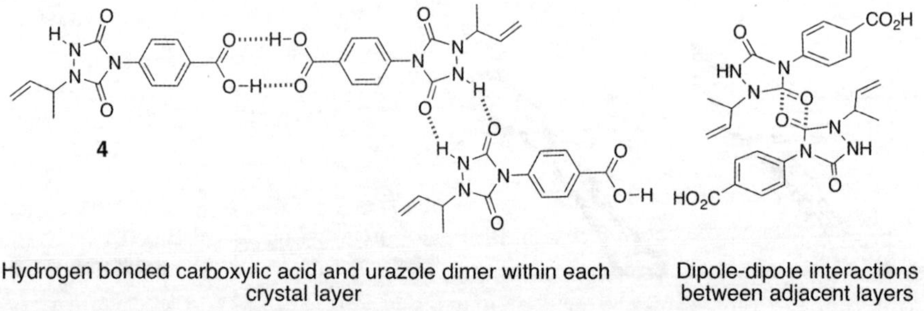

Hydrogen bonded carboxylic acid and urazole dimer within each crystal layer

Dipole-dipole interactions between adjacent layers

Figure 4.5   Supramolecular interactions (blue dashes) exhibited in the solid state by the urazole residue (**4**) developed by Stadler and co-workers.[38]

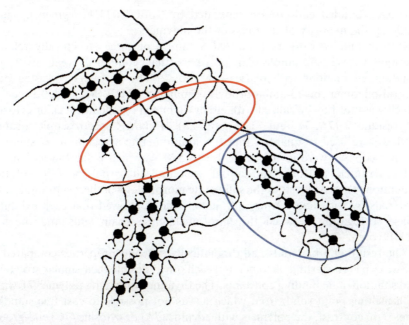

Figure 4.6 Schematic nanoscale structure of the urazole-functionalized poly(butadiene) produced by Stadler and co-workers. Urazole-rich hard domains and poly(butadiene)-rich soft domains are highlighted by the blue and red ellipses respectively.
Adapted with permission from reference [38].

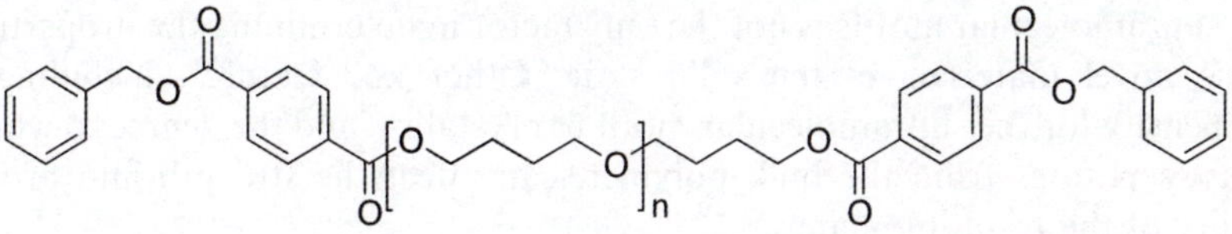

5

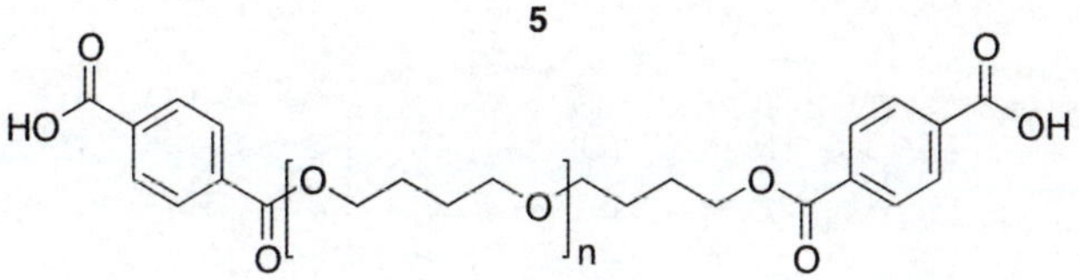

6

Figure 4.7 Structure of the end-functionalized pTHF polymers produced by Lillya and co-workers.[41]

50 to 70 °C, the acid-terminated p-THF (**5**) had a storage modulus that was two orders of magnitude greater than that of the benzyl terminated pTHF (**6**). This difference in thermomechanical properties was attributed to dimerization and subsequent crystallization of the acid residues. In agreement with the model proposed by Stadler, the morphology consisted of 'hard' domains, rich in

hydrogen bonded end-groups, separated by 'softer' pTHF segments, which made up the majority of the mass of the polymers.

Rowan and co-workers produced a family of three structurally related, thermosensitive, supramolecular polymers that combined weak hydrogen bonding interactions with phase separation.[45] The polymers consisted of a central oligomer, bis-(3-aminopropyl)-terminated pTHF ($M_n \approx 1400\,g\,mol^{-1}$), which was end-functionalized with either thymine (T), adenine (A), or cytosine (C) residues[46] (**7A**, **7C** and **7T**, Figure 4.8). ^{1}H NMR spectroscopic titration studies were used to quantify that the solution-state association constants of the nucleobases were in the range 1.5 to $5\,M^{-1}$, which is far lower than the value required to form substantial supramolecular polymers in solution [Equation (1)].[47] Analysis of the solid-state morphology of the 3 supramolecular materials by wide angle X-ray diffraction studies showed that each exhibited phase segregation between the hard nucleobase chain ends and the soft pTHF core.

The resulting 3 materials had dramatically different properties compared to the waxy pTHF starting oligomer and each other, despite their similar structures and solution-state binding constants. The thymine-containing polymer **7T** was a high melting point solid, from which it was not possible to cast free standing films.[48] In contrast, the polymers with adenine (**7A**) or cytosine (**7C**) end-groups could be cast into transparent, free standing films with melting points of over 125 °C. In addition, whilst the cytosine-terminated polymer was flexible, adenine-functionalized pTHF was brittle and prone to fracturing.

The select examples given in this section have highlighted the responsive nature of supramolecular materials that make them strong candidates for producing novel healable materials. The solution-state binding constants for the supramolecular motifs is not the only factor in determining the properties of these novel materials in the solid state. Other key features, including the propensity for the supramolecular motif to crystallize and the degree to which it phase-separates from the bulk polymer, can affect the strength and process-ability of the resulting material.[49]

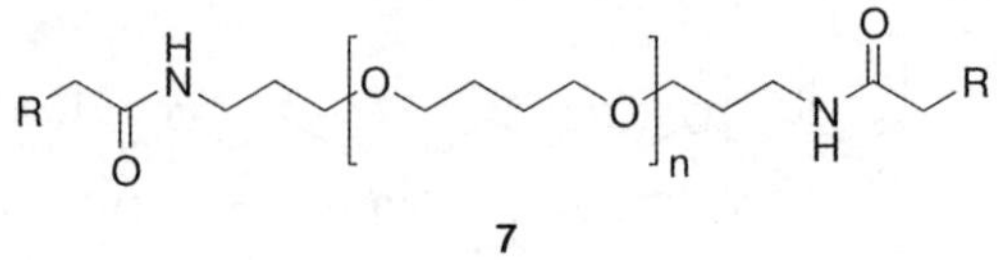

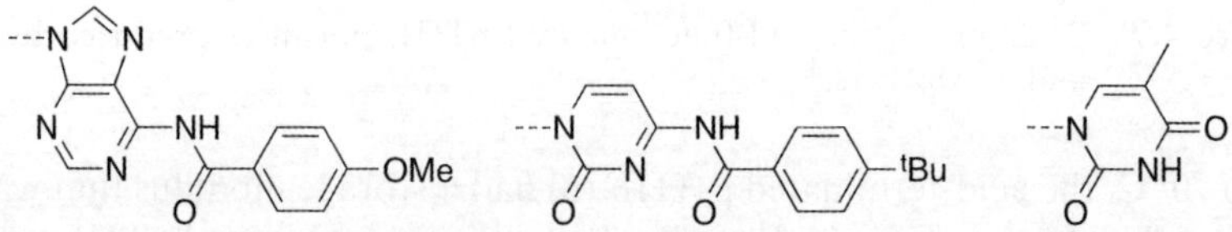

Figure 4.8 Structures of the thermoresponsive, nucleobase end-capped pTHF, studied by Rowan and co-workers.[43]

4.2.2 Healable Supramolecular Polymers Utilizing Hydrogen Bonding Interactions

From the responsive nature of the supramolecular systems described previously in this chapter, it is clear that under the correct conditions, many supramolecular polymers would be able to heal at least minor cracks by rearrangement of supramolecular interactions to bridge the fracture void. One of the first experimental indications that healable materials may be possible was published by Chino and Ashiura in 2001.[50] They produced a series of polyisoprenes, modified with varying loading levels of 1-amino-1,2,4-triazole (**8**, Figure 4.8) of up to 5.7 mol%. This functionalization procedure transformed the fluid poly(isoprene) into a thermoresponsive elastomer. Mechanical testing revealed that this new material could be reformed (recycled) at least 10 times without loss of key physical parameters, including elongation to breaking point and tensile strength. This suggests that during the recycling process, the covalent bonds remain intact and it is only the supramolecular bonds that are broken and reformed, a key requirement for a system if it is to withstand multiple break heal–cycles (see Section 1.3) (Figure 4.9).

In 2008, Leiber and co-workers produced the first weakly hydrogen bonded material that was designed, and experimentally verified, to undergo several break–heal cycles.[51] A scalable synthetic procedure utilizing inexpensive starting materials—including fatty acids (**9** and **10**), ethylene diamine and urea—was employed to produce a new hydrogen bonded polymer (Figure 4.10). Gel permeation chromatography of the product revealed it to be a complex mixture of branched oligomers with molecular weights of the order of 10^4 Da.[52]

The resulting polymer was a glassy solid at room temperature ($T_g = 28\,^\circ$C), but through plasticization by swelling in dodecane (11% w/w) was transformed into elastomeric material ($T_g = 8\,^\circ$C) which exhibited an elongation to break of 600%. When cut with a razor, the two new surfaces could be pressed back together by hand to regenerate a single bulk material that exhibited essentially identical properties to the pristine polymer. During these break–heal experiments, it was found that the single material had to be reformed quickly upon cutting in order to achieve the optimum healing efficiencies [Equation (1)]. Separation of the two cut surfaces for extended periods (up to 3 hours) resulted

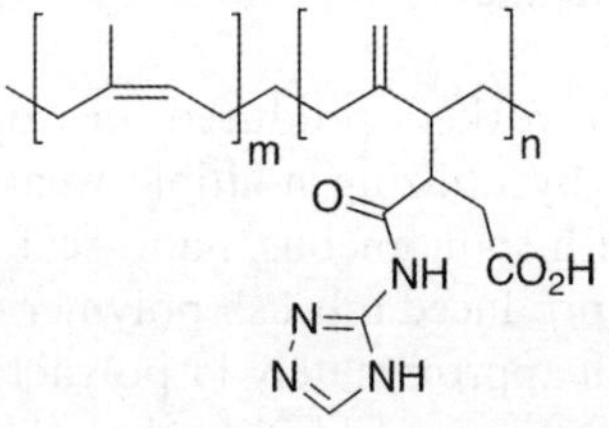

8

Figure 4.9 Structure of the recyclable polyisoprene produced by Chino and Ashiura (m-n = 100; n < 5.7).[46]

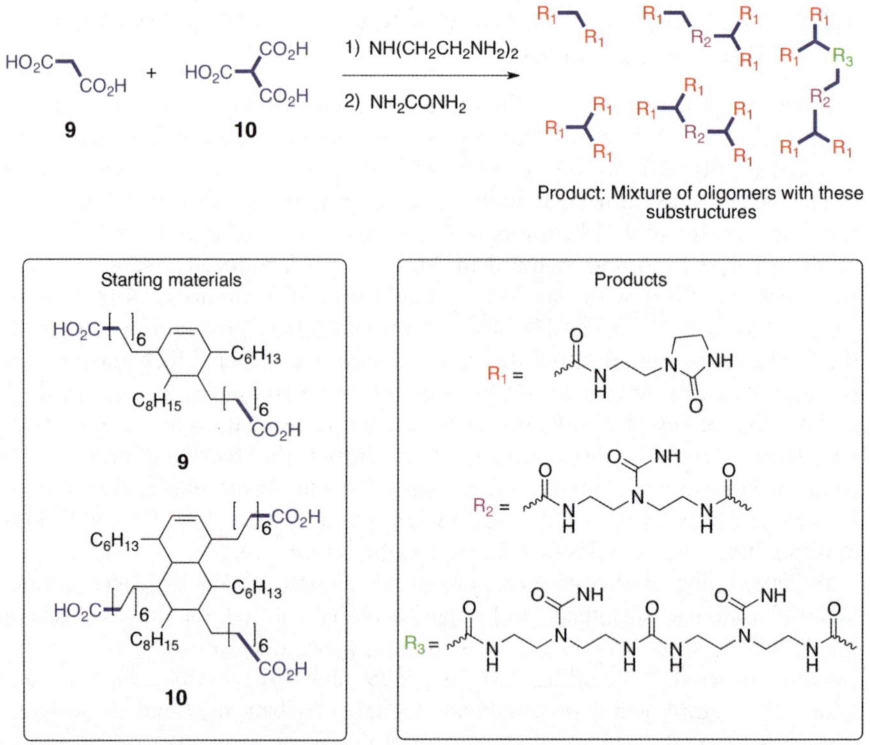

Figure 4.10   Synthesis of the branched polymer, produced by Leibler and co-workers, that exhibited self-healing properties after swelling in dodecane.[47]

in a dramatic reduction in healing efficiency. This result was rationalized by considering that the freshly revealed hydrogen bonding units at the surface of the material are able to rearrange with time to form new supramolecular interactions within the bulk of each cut section of bulk polymer. Thus, when two sections of fractured polymer are brought into contact after several hours of separation, the surfaces of the materials have reordered into a new thermodynamic minimum. In this thermodynamic equilibrium, any 'free' (unassociated) hydrogen bonding units that were abundant on the freshly cut surfaces are no longer available to bridge the fracture void and facilitate healing.

In 2012, Guan and co-workers produced an exquisitely designed healable supramolecular polymer by utilizing multiple weak hydrogen bonding inter- actions in conjunction with spontaneous, nano-scale ordering, driven by phase separation.[53] The group produced a brush polymer consisting of a polystyrene backbone (DP $\approx$ 114) with approximately 11 polyacrylic acid–amide side chains (each side chain DP $\approx$ 186, Figure 4.11a). In the solid state, immiscibility of the polystyrene and polar side groups drives the self-assembly of the polymer into a core-shell type structure. This results in an elastomeric solid material possessing a two-phase nano-structure, where hard-polystyrene domains are connected

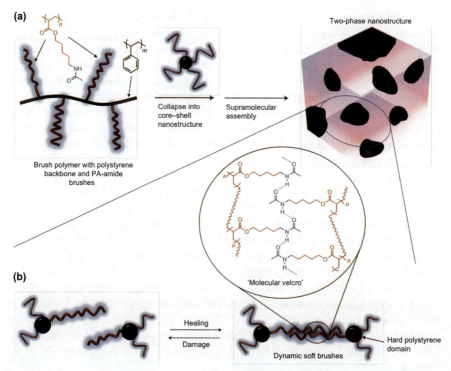

Figure 4.11   (a) Structure and schematic portrayal of the assembly of the supra-molecular, healable phase-separated brush copolymer produced by Guan and co-workers; b) Schematic of the proposed rupture and reforming of the supramolecular bonds formed between hydrogen bonding residues of the polymer brushes during a break–heal cycle.
Reprinted with permission from reference [49].

through a continuous phase of low T_g ($\sim 5\,^\circ$C), hydrogen bonded polyacrylic acid side chains. Healing studies were carried out by cutting pristine samples into two pieces and gently pressing (by hand) the separated faces together (Figure 4.11b). Comparison of the stress *versus* strain plots (Section 1.5) for the pristine and healed samples showed that healing efficiency increased with healing time and was as high as 92% after 24 hours contact time.

Control experiments, conducted on polymers produced by either blocking the hydrogen bonding amide residues or removing the polystyrene backbone, resulted in materials with at least a 10 fold decrease in tensile modulus when compared to the designed healing system.

4.3   Healable Supramolecular Polymers Utilizing π–π Stacking Interactions

Although supramolecular materials harnessing hydrogen bonding interactions are by far the most prevalent in the literature, and were the basis of the first thermoresponsive and healable materials, in principle, any supramolecular interaction may be exploited to similar effect. Indeed, interactions that do not depend on hydrogen bonding are less likely to be diminished by environmental contaminants such as moisture, which is unavoidable in a real-world setting.

The association between electronically complementary aromatic π-systems[23] has found widespread applications in supramolecular chemistry[54] since it was first reported in the early 1980s.[55,56] Colquhoun, Hayes and co-workers have recently harnessed these interactions to produce a family of healable supramolecular polymer blends.[57] The key design element in this work was the production of a readily accessible 'chain-folding'[58–60] motif comprised of two π-electron deficient naphthalene-1,4,5,8-tetracarboxylic diimine units, separated by a short diether linking unit (see complex A in Figure 4.12).

Computational modeling suggested that this configuration should enable the encapsulation of a π-electron rich aromatic residue, such as aspyrene, to produce a well defined π-stacked complex (Figure 4.12).[57] Furthermore, the introduction of additional complementary π-electron deficient and rich systems would clearly result in new complexes exhibiting association constants that increased in a highly predictable manner (see, for example, complexes A, B and C in Figure 4.12). Spectroscopic analysis of small molecule analogues of each of these computational models supported these predictions, and a family of related supramolecular complexes with binding constants that increased from 130 to 11000 M^{-1} were readily produced (Figure 4.12, complexes A to C, respectively).[57]

It was recognized that the key structural features of these supramolecular, chain-folded complexes could be rapidly introduced into polymeric systems. Thus a first generation material was produced; it consisted of an oligomeric polydiimide containing multiple chain-folding units, and was found to form a

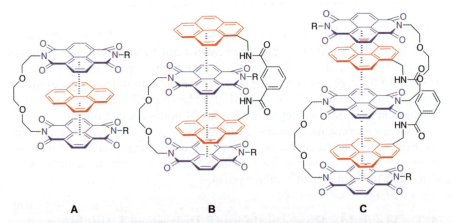

A   **B**   **C**

Figure 4.12   Structures of the complexes designed and modeled to produce a series of related supramolecular assemblies supported by π–π stacking interactions (R = Me). The complexes were synthetically verified (R = 2-ethyl hexane) to have binding constants of 130, 3500 and 11000 M^{-1} (A to C respectively).[53]

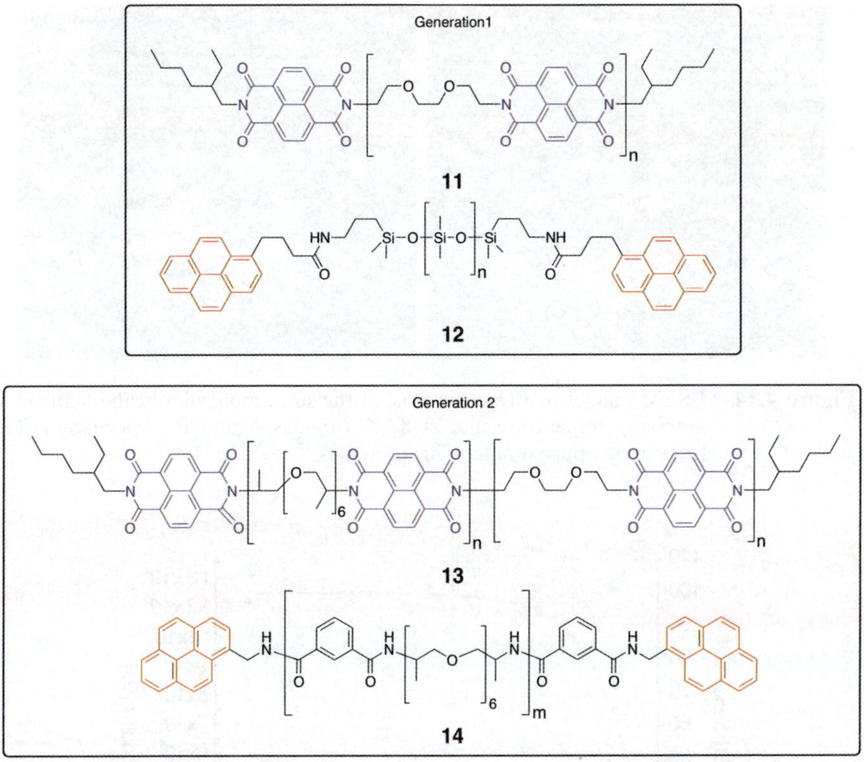

Figure 4.13 The structures of two generations of healable supramolecular polymers produced by Colquhoun, Hayes and co-workers.[57,58]

visually homogeneous blend with a pyrene end-capped polydimethylsiloxane (Figure 4.13, **11** and **12** respectively).[61] Although reported to be brittle, this supramolecular material was observed to undergo crack healing in response to a thermal stimulus.

By maintaining the π-electron deficient and π-electron rich residues that are present in each component of the blend, the main chains of the polymers could be structurally altered to produce a second generation of materials with improved physical properties (Figure 4.13, **13** and **14**). This was achieved by introducing an oligomeric diamine (Jeffamine®) into the system to produce two new polymers that, when blended, could be cast into free standing films.[62]

Healing of the second generation material was visualized by variable temperature environmental scanning electron microscopy (ESEM), which demonstrated that fractures of approximately 70 μm were rapidly (in less than a minute) eradicated by heating the material above 80 °C (Figure 4.14).

Quantitative data on the physical properties of the healed material were obtained by using a procedure that involved cutting the samples in two,

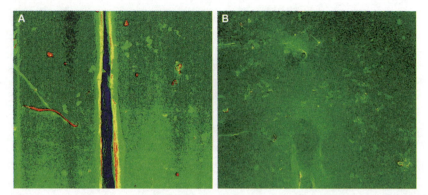

Figure 4.14   ESEM images of a fracture zone of the supramolecular healable blend at
ambient temperature and at 87 °C (images A and B respectively). The
fracture is approximately 70 μm across.[58]

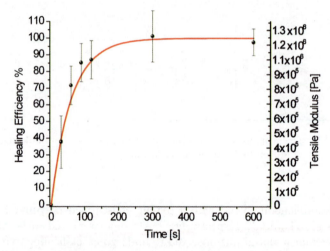

Figure 4.15   Plot of healing efficiency (and tensile modulus) as a function of healing
time at 50 °C for the supramolecular healable blend produced by
Colquhoun, Hayes and co-workers.
Reproduced with permission from reference [58].

overlapping the edges of the samples and keeping them in contact for varying
periods of time (up to 10 minutes) at 50 °C, prior to measuring the strength of
the sample. Comparison of the tensile modulus of the pristine sample with the
values measured after healing demonstrated that the material exhibited a
healing efficiency, η_{eff}, of 100% after a healing time of approximately 5 minutes
(Figure 4.15). A similarly high η_{eff} was recorded over 3 break–heal cycles, even
after the broken edges of the sample were separated for 24 hours before
subjecting the sample to the healing conditions.

A significant design feature of this healing polymer system is that the composition of the polymer backbone and the strength of the supramolecular interactions can be varied independently. This feature was further demonstrated by synthesizing a new polymer (**15**) that contained the same poly(amide) backbone as **14,** but differed in that it had dipyrene, tweezer-type end-groups (Figure 4.16).[63] These new end-groups were predicted to be able to form complexes akin to B and C (Figure 4.12). These complexes exhibit greater association constants than those formed in polymer blends where one component contains only a single pyrenyl residue at the chain ends (which can only interact by forming complex A, Figure 4.12).

Blends of dipyrene tweezer end-capped material **15** and chain folding polymer **13** exhibited a modulus of toughness of 3×10^8 Pa, an order of magnitude greater than that observed for a blend of **13** and **14**. This greatly enhanced strength was achieved at the cost of extended healing times and resulted in 100% η_{eff}, taking 160 minutes at 140 °C to recover the tensile modulus of the pristine material, compared to 5 minutes at 50 °C for the blend maintained by weaker supramolecular interactions. Thus there is a clear tradeoff between tailoring the association constant to produce strong materials and pushing the healing conditions outside acceptable parameters to induce healing.

Colquhoun, Hayes and co-workers also demonstrated that it was possible to produce healable materials that contained two distinct types of supramolecular interactions. To achieve this, a pyrenyl end-capped polyurethane (**16**, Figure 4.17) was produced through copolymerization of methyl diphenyl diisocyanate with dihydroxypolybutadiene.[47,64] This one-pot procedure delivered the target polymer, which featured both hydrogen bonding moieties (ureas and urethanes) and π-electron rich pyrenyl residues, in common with the previous healable polymer blends (Figures 4.13 and 4.16)

Blends of this hydrogen bonding polymer (**16**) and chain folding polymer **13** were tough, elastomeric products which could be stretched to 170% (*i.e.* a 2.7-fold increase in the starting dimension) of their initial length before breaking, compared to just 70% elongation for **13 + 15**. After one break–heal cycle, this supramolecular polymer blend retained 91% of its original elasticity and 70% of the pristine modulus of toughness, and the modulus of toughness remained relatively constant over 5 break–heal cycles.

This section has demonstrated that it is possible to produce healable materials that harness supramolecular interactions other than conventional hydrogen bonding motifs. Although these systems have been successfully and rationally optimized from a molecular level understanding, in common with most supramolecular materials, they are generally elastomers with glass transition temperatures well below room temperature. They are, therefore, unsuitable for many common uses of polymers, for example as lenses in spectacles or cups and containers, which must be fabricated from rigid, high modulus, materials. The following section details recent steps taken to rectify this problem.

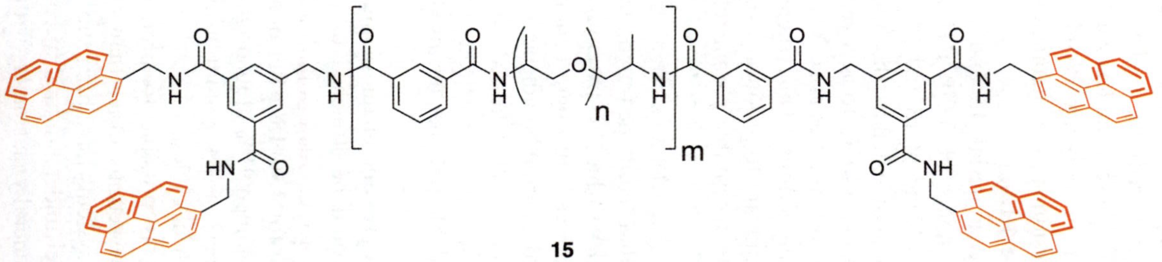

Figure 4.16 Structure of the dipyrene end-capped polymer, produced by Colquhoun and co-workers, that demonstrates the connection between association constant, mechanical properties and healing times.[59]

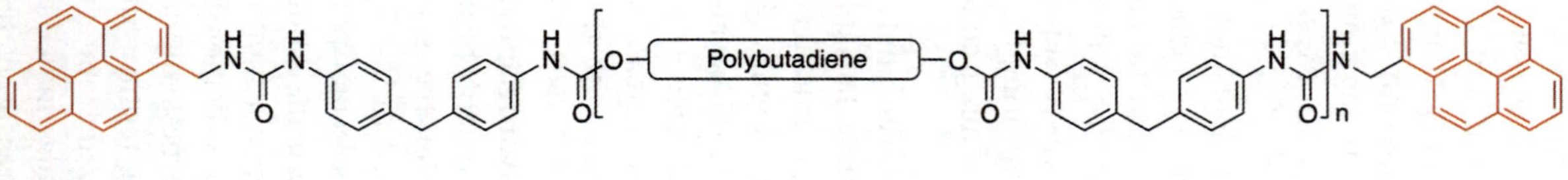

Figure 4.17 Structure of the pyrenyl end-capped polyurethane used in a two component supramolecular polymer blend.[60]

4.4 Healable Supramolecular Nano-composite Materials

Over the past 30 years, polymeric materials have found uses across a range of applications that were previously the domain of metals and their alloys; for example, in the chassis of racing cars and the fuselage of passenger planes (see Section 1.1). The expansion in the use of polymeric materials into these high performance areas has been facilitated by the introduction of polymer composites. In these materials, a high strength filler component, such as woven carbon fiber, is impregnated with the bulk (matrix) polymer to produce a single material with far more desirable mechanical properties than either of its constituent components. The strength of these materials is derived from the transfer of stress across the composite through the extremely tough filler component, rather than through the weaker polymer matrix. Therefore, in order to achieve healable composite systems, the mode of healing of both the matrix and the filler must be addressed.

In 2011, Fox *et al.*[65] tackled this problem by producing a composite material through impregnating the known healable polymer blend (**13 + 14**) (Figure 4.13) with cellulous nano-crystals (CNCs), which are a well studied[66] filler component for producing composites. CNCs can be extracted from many sources, including filter paper, cotton and wood; however, high aspect ratio CNCs can be processed from the bodies of small sea creatures known as *turnicates*. These CNCs measure around 5 nm in width and 0.5 μm in length and have a tensile strength of around 140 GPa. The high aspect ratio exhibited by the CNCs results in the formation of a continuous, percolating network that can transfer stresses imposed on the system through the filler loading levels (less than 10% w/w CNCs). During testing, the stress is transferred through the network of CNCs and across the macroscopic length scale of the sample. Thus, unlike in a conventional, woven carbon fiber composite (where breaking the fibers leads to an irrecoverable loss of integrity of the component), provided that there is sufficient energy to re-establish the continuous network of CNC, healing can be achieved.

Fox *et al.* studied the effect of varying the loading level of CNCs from 0 to 20 wt% filler in the healable supramolecular polymer blend (**13 + 14**). The tensile modulus of the material increased from 8 MPa to 261 MPa as the proportion of filler increased from 0 to 10 wt%. Above this loading level, the materials became weaker as a consequence of phase separation of the filler and matrix polymer, producing a non-homogeneous dispersion of CNCs, where some areas had extremely low loading levels of CNCs. In this situation, the strength of the material is reduced to that of the weakest section (area with the lowest concentration of CNCs) of the bulk material.

Healing experiments were conducted by producing dogbone-type samples (see Section 1.5), which were cut with a razor blade then positioned with the two cut edges overlapping and heated at 85 °C for 2 minutes; their mechanical properties were measured by dynamic mechanical thermal analysis (DMTA, see Section 1.5). Comparison of the strength of the healed samples to that of

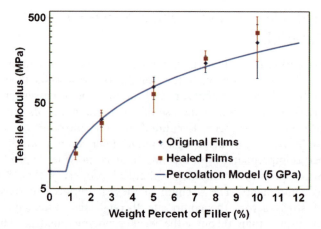

Figure 4.18   Plot of variation of the strength of the pristine and healed supra-molecular nano-composite materials as a function of wt% CNC filler. The blue line represents the predicted strength of the materials, as calculated by the percolation model.[64–66]
Reproduced with permission from reference [62].

pristine samples revealed healing efficiencies of at least 90% for all CNC loading levels of up to 10 wt% (Figure 4.18). In addition, it was found that all samples, both pristine and healed, exhibited tensile moduli in close agreement with that predicted from the percolation model,[67–69] suggesting that in all cases the CNCs remained homogeneously dispersed throughout the healable supra-molecular nano-composite.

This section has detailed the production and optimization of supramolecular healable polymers that utilize π–π stacking interactions. These materials were designed from the molecular length-scale perspective, which allowed the supramolecular interactions and covalent polymeric components of the material to be studied and improved upon separately, resulting in increasingly tough polymers. Including a nano-filler component in the polymer blend demonstrated a new methodology for dramatically improving the performance of these materials whilst retaining the healable characteristics and may find widespread application in future research.

The following section describes the application of metal–ligand interactions in producing healable materials.

4.5   Healable Supramolecular Polymers Utilizing Metal–Ligand Interactions

Metal–ligand interactions have been widely studied for the production of responsive supramolecular materials. For example, materials harnessing these interactions have been shown to exhibit thixotropic and sensing properties, amongst others.[70]

The first example of using metal–ligand interactions to produce healable materials was reported by Rowan, Weder and co-workers, who studied metallo-supramolecular polymers comprising a low molecular weight ditopic macromonomer along with transition or lanthanide metal salts. A unique feature of these materials is that they were able to undergo crack/scratch healing upon exposure to UV light, rather than direct thermal stimulation.[71,72]

In these studies, a low T_g telechelic poly(ethylene-*co*-butylene) macro-monomer (**17**) with 2,6-bis(1'-methylbenzimidazolyl)-pyridine (Mebip) ligand chain ends was combined with $Zn(NTf_2)_2$ and/or $La(NTf_2)_3$ (Figure 4.19a) to produce flexible elastomeric films of **17**·[$Zn(NTf_2)_2$] and **15**·[$La(NTf_2)_3$].[73] Characterization of these supramolecular polymers using small-angle X-ray scattering and transmission electron microscopy revealed microphase-separated lamellar morphologies in which the metal–ligand complexes formed a "hard phase" that physically crosslinks "soft" domains formed by the poly(ethylene-*co*-butylene) cores (Figure 4.19b), in a manner similar to several hydrogen bonded supramolecular polymers that feature analogous cores. This structure is the origin of the material's intriguing mechanical properties. The metal–ligand bonds are dynamic in nature and thus serve as the supramolecular motif that can be used to polymerize/depolymerize the materials on demand. The Mebip ligand and its Mebip–metal complexes exhibit an absorption band with maxima at *ca.* 310 nm and 340 nm and are non-absorbing above 390 nm. Thus the materials are essentially colorless and can be excited in the near UV regime. As a result of the low fluorescence quantum yield, the absorbed optical energy is converted into heat. This causes temporary disengagement of the metal–ligand complexes and a concomitant reversible decrease of the metal-losupramolecular polymers' molecular weight and viscosity, thereby allowing rapid and efficient healing. Given that the polymers are not very good thermal conductors, the photo-thermal conversion process results in selective heating of the material around areas that are directly exposed to the light. Thus when deliberately damaged, samples of **17**·[$Zn(NTf_2)_2$] and **17**·[$La(NTf_2)_3$] were exposed to UV irradiation (320–390 nm and an intensity of 950 mWcm^{-2}) an exposure time of < 1 minute was sufficient to completely heal the defects without degrading the sample (Figure 4.19c). As expected, only the area within the light beam was healed (white dotted circle in Figure 4.19c). Monitoring the light-induced healing process with an IR camera confirmed that the areas of the films exposed to light could reach temperatures of *ca.* 200 °C in that time frame. Mechanical analysis of fully healed samples revealed that the original mechanical properties of these materials can be re-established. Furthermore, these systems also have the ability to be repeatedly damaged and re-healed. Interestingly, the nature of the metal ion used to assemble the macromonomer (**17**) was found to play a significant role in the efficiency of the photo-healing process. La^{3+}/Mebip complexes are more labile and dissociate at lower temperatures than Zn^{2+}/Mebip complexes,[74–76] and it was found that under the same healing conditions (2×30 seconds of UV exposure), the **17**·[$La(NTf_2)_3$] films healed more efficiently than the **17**·[$Zn(NTf_2)_2$] films. This is consistent with the much lower stability (and higher level of light-induced depolymerization) of La^{3+}/Mebip complexes *vis a vis* the Zn^{2+}/Mebip-based counterparts. Good healability could be imparted to the polymers by changing

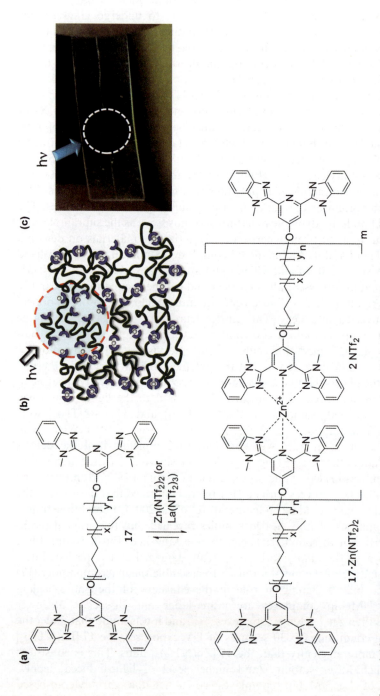

Figure 4.19 (a) Chemical structures of macromonomer **17** and the metallosupramolecular polymer **17**·[Zn(NTf2)2]; (b) Schematic representation of the lamellar structure of **17**·[Zn(NTf2)2] and the light-activated healing of metallosupramolecular polymers; (c) Image illustrating the optical healing of a film of **17**·[Zn(NTf2)2] with UV light ($\lambda = 320$–$390\,nm$, $950\,mWcm^{-2}$, 2×30 seconds). The white circle identifies the area where a light beam was locally applied to remove the deliberately introduced cut running through the sample.
Figure adapted with permission from references [68, 69].

the Zn^{2+} : Mebip stoichiometry to 0.7:1, suggesting that the dynamics of the light-induced depolymerization, and thereby the healing behavior, can be strongly influenced by the presence of an excess of free ligands. The concept of photo-thermal induced healing of supramolecular materials appears to be applicable to any supramolecular polymer with a binding motif that is sufficiently dynamic. The ability to change the light-absorbing motif allows one to tailor the wavelength of the light that is used for healing. The combination of this new approach with an additional mechanochromic response[77,78] promises access to autonomously functioning, self-healing materials, in which light is only absorbed at defect sites.

4.6   Introduction to Healable Supramolecular Gels

Supramolecular gels have become a prominent area of research over the last 15 years, with particular interest in optimizing their use for optical, electron and biomedical applications.[79] These gels are formed by the self-assembly of the gelator into a sample spanning network that sequesters the solvent through capillary action. Importantly, the gelator to solvent ratio is very low, typically in the range 0.5 to 10 wt%. The effect of this process is easy to visualize and is frequently demonstrated using the classic *tube inversion test* (Figure 4.20). During this experiment, a sample tube containing the gel is turned upside down and the gel is deemed to have formed successfully if it can resist the force of gravity.

Supramolecular gels are further separated by the structures of the gelating molecules: *polymeric gelators* consist of high molecular weight building blocks,[80] whereas *low molecular weight gelators* are typically small molecule single compounds measuring a few hundred Daltons in weight.[81] Furthermore, each class of gelator usually only forms gels in either organic or aqueous solvents and are thus termed *organogelators* and *hydrogelators* respectively.

As a consequence of the high solvent content of gels, they are generally weak materials (G' less than a few kPa, see Section 1.5.3), unsuitable for structural

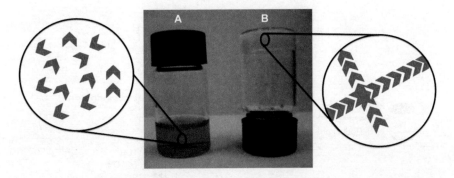

Figure 4.20   Photographs of a supramolecular gelator in solution (vial A) and gelled state during a tube inversion test (vial B). The schematics show the supramolecular gelator (blue arrows) in solution and after self-assembly (left and right respectively).

applications. However, in certain, more niche circumstances, for example, to stabilize the electrolyte in batteries or as tissue engineering scaffolds, the ability of these materials to heal fractures and defects could be hugely beneficial.

By definition, supramolecular gels contain similar interactions to those found in the materials described in Sections 4.1 to 4.5, and healing is therefore accomplished by re-engagement of the supramolecular interactions across the fracture void. Indeed, this procedure can be facilitated in gels by fast molecular level dynamics, as a consequence of the ready availability of solvent. Sections 4.6.1 and 4.6.2 use select examples to show the diversity of chemistry that has been used to synthesize supramolecular healable polymeric gels and low molecular weight gelators respectively.

4.6.1   Healable Supramolecular Polymeric Gels

In 2011, Harada and co-workers produced an interesting example of a redox healable polymeric hydrogel by exploiting the host–guest interactions between ferrocene and β-cyclodextrin.[82] In its native state, ferrocene is known to form a strongly-bound inclusion complex with β-cyclodextrin, whereas oxidation to produce the ferrocenylcation results in decomplexation of the system (Figure 4.21). The healable material comprised two structurally related polymers based on a poly(acrylic acid) ($M_w = 25$ kDa) backbone. One polymer contained approximately 5 mol% β-cyclodextrin, whereas the second was substituted with approximately 2.7 mol% ferrocenyl residues. A mixture of these two polymers produced a stable hydrogel at 2 wt% concentration. Chemical (NaClO) or electrically induced oxidation of the ferrocenyl side groups resulted in a loss of integrity of the gel as a consequence of the reduction in strength of the host–guest complex. The process was reversed by reduction of the ferrocenylcation back to the neutral form (either chemically or electronically).

Healing experiments were conducted by cutting the hydrogel and placing the freshly-cut edges in contact with each other for 24 hours at room temperature;

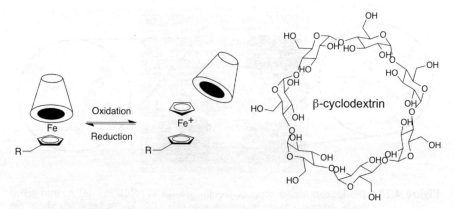

Figure 4.21   Schematic of the redox-induced reversible complexation of β-cyclo-dextrin with ferrocenyl derivatives.[79]

during this time, the two sections fused to reform a single hydrogel with approximately 84% healing efficiency (Figure 4.22). Control experiments demonstrated that addition of a chemical oxidant or a competitive chemical guest for the β-cyclodextrin to the cut surface of the hydrogel resulted in a loss of healing ability of the system as a consequence of the disruption of the host–guest interactions.

In 2012, Phadke *et al.* produced a lightly crosslinked polymer hydrogel and demonstrated its use in producing healable, chemically resistant coatings and tissue engineering scaffolds, as well as in drug delivery applications.[83] The gel comprised a loosely covalently crosslinked acrylonitrile network with multiple carboxylic acid side groups (Figure 4.23a). Typically, crosslinked polymeric gels would not show healing properties unless dynamic covalent bonds capable

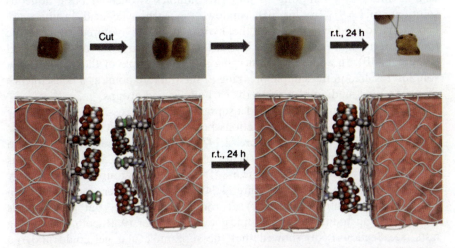

Figure 4.22   Example of a healing experiment where two severed pieces of hydrogel were placed in contact with each other for 24 hours and fused. The schematic shows the proposed reversible engagement of the ferrocenyl (iron atoms in green) and β-cyclodextrin residues during healing. Adapted with permission from reference [79].

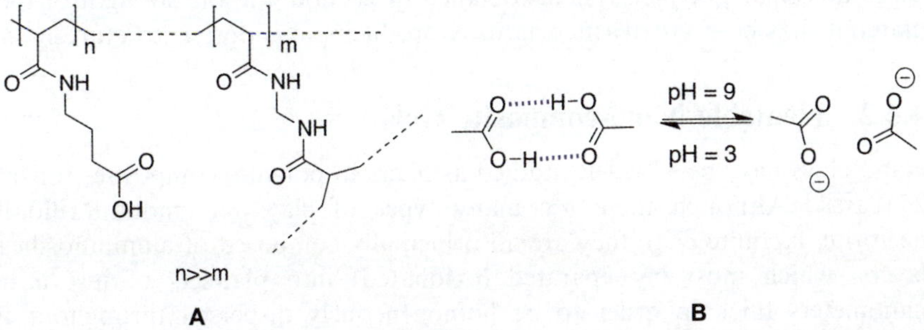

Figure 4.23   (A) Chemical structure of the healable hydrogel possessing both supramolecular and chemical crosslinks; (B) Scheme showing the proposed pH switchable supramolecular interactions (blue dashes) required to promote healing.

of forming new covalent bonds that bridge the fractured surfaces are present (see Section 3.3). In this example, at pH = 3, the pendent acid side groups are protonated and can form additional supramolecular crosslinks through hydrogen bonding interactions. Thus, at low pH, placing two pieces of this gel in contact with each other, even for a few seconds, results in the two sections of gel becoming fused, although increased contact time (24 hours) results in stronger joints. The healing efficiency of this system was modest (66%) compared to most supramolecular systems, an observation that is consistent with the loss of the continuous covalent network throughout the sample, and the healed fracture being supported by hydrogen bonding interactions only.

The role that hydrogen bonding plays in the healing process was elegantly demonstrated by attempting to join two samples swollen in basic aqueous solutions (pH = 9). In this pH window, the carboxylic acid residues are deprotonated and cannot from the hydrogen bonds required to heal the fracture, resulting in the complete loss of the healing properties of the hydrogel (Figure 4.23B). In addition, submerging a healed sample of the gel in a urea solution resulted in the sample breaking at the fracture point as a consequence of competitive binding of urea to the carboxylic acid groups.

Huang and co-workers produced a supramolecularly crosslinked hydrogel by employing the well known interactions between quaternary ammonium salts and crown ether residues (Figure 4.24A).[84] The two component system comprised a poly(methyl methacrylate) backbone ($M_n = 9.5$ kDa) which was functionalized with dibenzo[24]crown-8 (DB24C8) side groups (approximately 7 per polymer chain) (**18**, Figure 4.24B). This polymer was soluble in chloroform/acetonitrile (50:50 v/v) at room temperature, but formed gels immediately on addition of the ditopic guest crosslinker **19**. Rheological healing tests (see Section 1.5.3) showed that the supramolecular gel could undergo multiple break–heal cycles with typical healing efficiencies of greater than 95%.

The examples described in this section demonstrate the diverse chemistries that have been employed to produce supramolecular healable gels in this young, but growing field of research. Clearly, the low strength of gels restricts their use in specialized applications. However, as in the case of the healable supramolecular polymer systems detailed in Section 4.5, the strength of these materials has been greatly increased by producing composite structures.

4.6.2 Healable Nano-composite Gels

Nano-clays have been widely studied as fillers in polymer composites for over 25 years.[85] Although there are many types of clay (*e.g.* montmorillonite, hectorite, laponite *etc.*), they are all principally composed of aluminnosilicate layers, which must be separated (exfoliated) into platelets of just a few nanometers thick in order to be homogeneously dispersed throughout the polymer matrix. In a series of papers beginning in 2002, Haraguchi and co-workers reported that it was possible to harness the same nano-clay fillers to increase the strength of conventional covalent hydrogel polymers, [for example, poly(*N*-isopropylacrylamide)].[86–89]

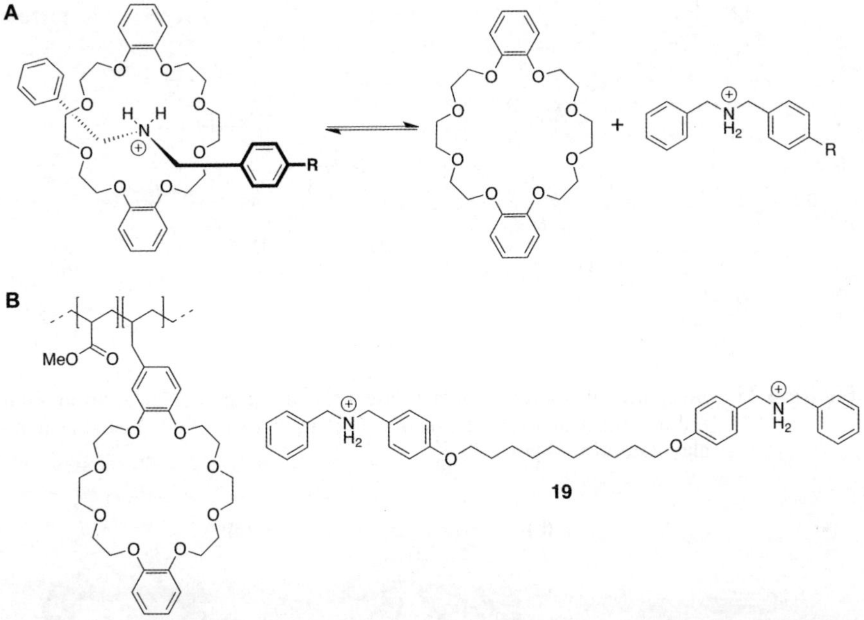

Figure 4.24 (A) Reversible formation of a crown ether/ammonium ion pseudoro-taxane; (B) Structures of the crown ether functionalized poly(methyl methacrylate) **18** and ditopic ammonium ion crosslinker **19** used by Huang and co-workers to form supramolecular healable gels.[81]

The first demonstration that nano-composite gels can be healed was reported by Aida and co-workers in 2010.[90] The group produced a multicomponent hydrogel consisting of clay nano-sheets, sodium polyacrylate (**20**) and guarnidium functionalized dendrimers (**21**) of varying generations in water (Figure 4.25).

The organic components were designed for separate tasks, the linear polymer **20** performing a key role in exfoliating the nano-clay through favorable interactions with the polar groups on the surface of the clay platelets (Figure 4.26b). The resulting polymer-coated nano-clay was then crosslinked through transient salt bridges formed between the positively charged guar-nidium residues on the dendrimers and the acid groups on linear polyacrylate (Figure 4.26c), generating a three-dimensional supramolecular network. The strength of the hydrogels was found to increase with increasing number of surface groups on the dendrimers (by increase in generation number from 1 to 3), with the strongest gels exhibiting a storage modulus of approximately 0.5 MPa (see Section 1.5.3).

The healing ability of the nano-composite hydrogel was demonstrated by rheological experiments (see Section 1.5) and visually by placing cubes of gel in contact with each other to form a self-supporting structure (Figure 4.27). In common with the supramolecular rubber reported by Leibler and co-workers[51] (Section 4.2.2), separation of cut sections (for more than a minute) resulted in

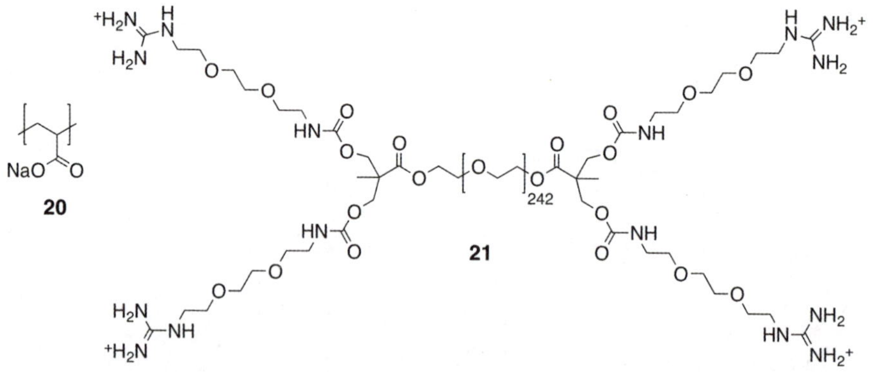

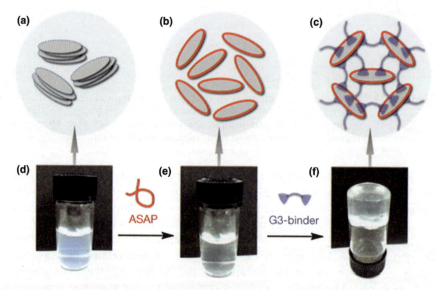

Figure 4.25 Structure of sodium polyacrylate (**21**) and the first generation guar-
nidium functionalized dendrimer (**22**) used to produce healable nano-
clay composite hydrogels.[91]

Figure 4.26 Schematic representation of undispersed nano-clay platelets, exfoliated
sodium polyacrylate (ASAP) coated nano-clay platelets and the
dendrimer crosslinked supramolecular gel (a to c respectively).
Photographs of aqueous samples of undispersed nano-clay platelets,
exfoliated sodium polyacrylate (ASAP) coated nanoclay platelets and the
dendrimer crosslinked supramolecular gel (d to f, respectively).
Adapted with permission from reference [91].

loss of healable characteristics as a consequence of the rapid rearrangement of
the supramolecular motifs from the cleaved surface into the bulk over short
time periods.

More recently, Haraguchi and co-workers also used nano-clay as the filler
component to construct two healable composite hydrogels which differed

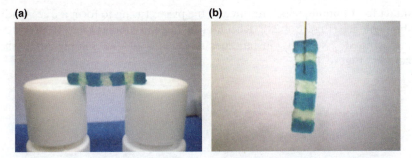

Figure 4.27 Photographs of a 'healed' sample of free standing nano-clay hydrogel formed by fusing 7 sections of freshly cut gel (4 sections dyed blue for clarity). Reproduced with permission from reference [91].

according to the structure of the polymeric components.[91] These systems were produced by free radical polymerization of either *N,N*-dimethylacrylamide or *N*-isopropylacrylamide in water containing varying quantities of nano-clay, to generate free-standing gels of approximately 10–15 wt% total solid content. Visually impressive healing of deep cuts in the gels was observed by leaving the samples at 37 °C for 48 hours (Figure 4.28). Further studies showed that healing efficiency was related to the contact time of the severed edges, with η_{eff} of 27%, 57% and almost 100% observed after healing times of 1, 4 and 10 hours respectively (at 50 °C in a sealed container). Unusually for supramolecular systems, healing was observed even after the severed surfaces had been separated for several days. In addition, sections of hydrogels composed of different polymers—(poly(*N,N*-dimethylacrylamide) or poly (*N*-isopropylacrylamide)—were able to be fused into freestanding structures that could be manually stretched and compressed without breaking.

4.6.3 Healable Molecular Gels

Although there are several reports of healable, low molecular weight gelators, the mode of healing is frequently through dynamic covalent chemistries, rather than supramolecular interactions,[92,93] and are therefore beyond the scope of this chapter. In healable molecular gels, the large number of supramolecular interactions, combined with the high solvent content, allows easy diffusion of the gelator molecules, accounting for the thixotropic and shear thinning behavior frequently exhibited by gels. Although all molecular gels will heal given the appropriate stimuli, a this was recently demonstrated by Dastidar and co-workers,[94] who used the supramolecular synthon approach[95] to guide the design of structurally simple organogelators derived from ammonium salts of amino acids derivatives. One of these gels, derived from Boc-protected glycine (**22**) and dibenzyl amine (**23**) (7 wt% in nitrobenzene) was observed to form freestanding gels that were stable for many months at ambient conditions (Figure 4.28). Healing was visually demonstrated by placing five cubes of gel in contact for 1 minute, which resulted in the pieces fusing to form a free standing cuboid of gel (Figure 4.28). Rheological healing studies (see Section 1.5.3) demonstrated essentially instantaneous healing of the sample after shear fractures (Figure 4.29).

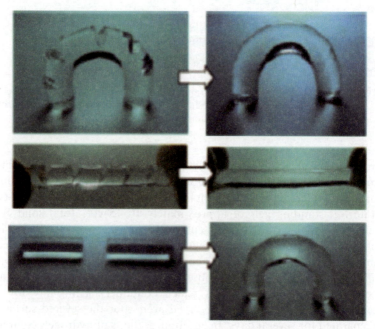

Figure 4.28 Examples of healing deep cuts and completely severed samples of nano-clay hydrogels.
Reproduced with permission from reference [92].

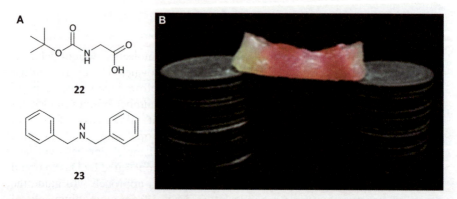

Figure 4.29 (A) Structures of Boc-protected glycine (**20**) and dibenzyl amine (**21**) used to form freestanding organogels; (B) Example of a self-supporting sample of a healed gel produced from joining 5 pieces of gel (2 sections dyed pink for clarity).
Adapted with permission from reference [84].

4.7 Conclusions and Future Perspectives

This chapter has used examples from the literature to demonstrate the key design criteria that must be fulfilled in order to produce healable supramolecular systems. Although the strength and directionality of the supramolecular interactions harnessed within each of these materials is of great importance, in the solid state, crystallinity and phase separation of components can also be harnessed to build in additional strength and functionality.

It is proposed that in supramolecular systems, fractures occur through scission of the weaker, supramolecular interactions within the materials, rather than the stronger covalent bonds. Healing is facilitated by re-engagement of the supramolecular motifs across the fracture void. However, these universal design elements may also impose an upper limit on the strength and scope of application of some of these materials, for example:

i) Materials using supramolecular motifs may always fail before conventional covalently crosslinked materials as a consequence of their lower bond energies.
ii) In order to allow enough molecular motion for healing to occur on a reasonable time scale, the phase of the material containing the supramolecular motif must be at a temperature greater than its T_g (or T_m), or in the gel phase.

These requirements result in the high proportion of weak, elastomeric or gel-type supramolecular materials detailed in this chapter. Increasing the strength of the supramolecular interactions has been demonstrated to improve mechanical properties, but this comes with the requirement for harsher healing conditions (Section 4.3). Whilst healable supramolecular nano-composites may offer a path to greatly enhanced strength (Sections 4.4 and 4.6.2), their use has only been demonstrated in highly academic scenarios that are far from real-world applications.

In conclusion, the ground-work in setting out the design, synthesis and evaluation protocols of healing supramolecular materials has been completed. However, major challenges lie ahead—specifically in producing materials that exhibit high strength and high glass transition temperatures—in moving the field from a purely academic discipline to a situation whereby supramolecular polymers provide solutions for real problems on an industrial scale.

References

1. S. D. Bergman and F. Wudl, *J. Mater. Chem.*, 2008, **18**, 41.
2. S. Burattini, B. W. Greenland, D. Chappell, H. M. Colquhoun and W. Hayes, *Chem. Soc. Rev.*, 2010, **39**, 1973.
3. E. B. Murphy and F. Wudl, *Prog. Polym. Sci.*, 2010, **35**, 223.
4. M. W. Urban, *Prog. Polym. Sci.*, 2009, **34**, 679.
5. B. J. Blaiszik, S. L. B. Kramer, S. C. Olugebefola, J. S. Moore, N. R. Sottos and S. R. White, *Annu. Rev. Mater. Res.*, 2010, **40**, 179.

6. R. P. Wool, *Soft Matter*, 2008, **4**, 400.
7. S. Seiffert and J. Sprakel, *Chem. Soc. Rev.*, 2012, **41**, 909.
8. D. Y. Wu, S. Meure and D. Solomon, *Prog. Polym. Sci.*, 2008, **33**, 479.
9. J. A. Syrett, C. R. Becer and D. M. Haddleton, *Polym. Chem.*, 2010, **1**, 978.
10. K. Jud, H. H. Kausch and J. G. Williams, *J. Mater. Sci.*, 1981, **16**, 204.
11. Y. H. Kim and R. P. Wool, *Macromolecules*, 1983, **16**, 1115.
12. A. P. Esser-Kahn, S. A. Odom, N. R. Sottos, S. R. White and J. S. Moore, *Macromolecules*, 2011, **44**, 5539.
13. S. R. White, N. R. Sottos, P. H. Geubelle, J. S. Moore, M. R. Kessler, S. R. Sriram, E. N. Brown and S. Viswanathan, *Nature*, 2001, **409**, 794.
14. K. S. Toohey, N. R. Sottos, J. A. Lewis, J. S. Moore and S. R. White, *Nature Mater.*, 2007, **6**, 581.
15. See for example: A. P. Esser-Kahn, P. R. Thakre, H. F. Dong, J. F. Patrick, V. K. Vlasko-Vlasov, N. R. Sottos, J. S. Moore and S. R. White, *Adv. Mater.*, 2011, **23**, 3654.
16. G. M. L. van Germert, J. W. Peeters, S. H. M. Söntjens, H. M. Janssen and A. W. Bosman, *Macromol. Chem. Phys.*, 2012, **213**, 234.
17. N. K. Guimard, K. K. Oehlenschlaeger, J. Zhou, S. Hilf, F. G. Schmidt and C. Barner-Kowollik, *Macromol. Chem. Phys.*, 2012, **213**, 131.
18. L. Brunsvel, B. J. B. Folmer, E. W. Meijer and R. P. Sijbesma, *Chem. Rev.*, 2001, **101**, 4071.
19. *Supramolecular Polymers*, ed. A. Ciferri, Marcel Dekker, New York, 2000.
20. R. P. Sijbesma, F. H. Beijer, L. Brunsveld, B. J. B. Folmer, J. H. K. K. Hirschberg, R. F. M. Langer, J. K. L. Loweand and E. W. Meijer, *Science*, 1997, **278**, 1601.
21. T. Park, S. C. Zimmerman and S. Nakashima, *J. Am. Chem. Soc.*, 2005, **127**, 6520.
22. C. Fouquey, J-M. Lehn and A-M. Levelut, *Adv. Mater.*, 1990, **2**, 254.
23. C. A. Hunter and J. K. M. Sanders, *J. Am. Chem. Soc.*, 1990, **112**, 5525.
24. For a recent review see: V. A. Friese and D. G. Kurth, *Coord. Chem. Rev.*, 2008, **252**, 199.
25. A selected early paper: P. L. Anelli, P. R. Ashton, R. Ballardini, V. Balzani, M. Delgado, M. Teresa Gandolfi, T. T. Goodnow, A. E. Kaifer, D. Philp, M. Pietraszkiewicz, L. Prodi, M. V. Reddington, A. M. Z. Slawin, N. Spencer, J. F. Stoddart, C. Vincent and D. J. Williams, *J. Am. Chem. Soc.*, 1992, **114**, 218.
26. For an account of the development towards applications of these interactions, see: J. F. Stoddart and H. M. Colquhoun, *Tetrahedron*, 2008, **64**, 8231
27. For a review, see: E. R. Kay, D. A. Leigh and F. Zerbetto, *Angew. Chem., Int. Ed.*, 2007, **46**, 72.
28. S. J. Cantrill, G. J. Youn, J. F. Stoddart and D. J. Williams, *J. Org. Chem.*, 2001, **66**, 6587.
29. D. Coa, C. Wang, M. A. Giesener, Z. Liu and J. F. Stoddart, *Chem. Commun.*, 2012, **48**, 6791.

30. F. H. Huang and H. W. Gibson, *J. Am. Chem. Soc.*, 2004, **126**, 14738.
31. J-M. Lehn, *Angew. Chem. Int. Ed.*, 1988, **100**, 91.
32. C. J. Pederson, *Angew. Chem. Int. Ed.*, 1988, **100**, 1009.
33. D. J. Cram, *Angew. Chem. Int. Ed.*, 1988, **100**, 1021.
34. M. Barboiu, G. Vaughan, R. Graff and J-M. Lehn, *J. Am. Chem. Soc.*, 2003, **125**, 10257.
35. For a review of supramolecular polymerizations, see: J. D. Fox and S. J. Rowan, *Macromolecules*, 2009, **42**, 6823
36. For a review of the Suprapolix® material, see: A. W. Bosman, R. P. Sijbesma and E. W. Meijer, *Mater. Today*, April 2004, 34.
37. L. L. de Lucca Freitas and R. Stadler, *Macromolecules*, 1987, **20**, 2478.
38. C. Hilger and R. Stadler, *Macromolecules*, 1990, **23**, 2097.
39. C. Hilger, R. Stadler and L. L. de Lucca Freitas, *Polymer*, 1990, **31**, 818.
40. C. Hilger and R. Stadler, *Polymer*, 1991, **32**, 3244.
41. C. Hilger, M. Dräger and R. Stadler, *Macromolecules*, 1992, **25**, 2498.
42. C. Hilger and R. Stadler, *Macromolecules*, 1992, **25**, 6670.
43. M. Müller, U. Seidel and R. Stadler, *Polymer*, 1995, **36**, 3143.
44. C. P. Lillya, R. J. Baker, S. Hütte, H. H. Winter, Y-G. Lin, J. Shi, L. C. Dickinson and J. C. W. Chien, *Macromolecules*, 1992, **25**, 2076.
45. S. Sivakova, D. A. Bohnsack, M. E. Mackay, P. Suwanmala and S. J. Rowan, *J. Am. Chem. Soc.*, 2005, **127**, 18202.
46. For a review of nucleobases in supramolecular chemistry, see: S. Sivakova and S. J. Rowan, *Chem. Soc. Rev.*, 2005, **34**, 9.
47. P. J. Woodward, D. H. Merino, B. W. Greenland, I. W. Hamley, Z. Light, A. T. Slark and W. Hayes, *Macromolecules*, 2010, **43**, 2512.
48. S. J. Rowan, P. Suwanmala and S. Sivakova, *J. Polym. Sci., Part A: Polym. Chem.*, 2003, **41**, 3589.
49. J. Cortese, C. Soulié-Ziakovic, S. Tencé-Girault and L. Leibler, *J. Am. Chem. Soc.*, 2012, **134**, 3671.
50. K. Chino and M. Ashiura, *Macromolecules*, 2001, **34**, 9201.
51. P. Cordier, F. Tournilhac, C. Soulié-Ziakovic and L. Leibler, *Nature*, 2008, **451**, 977.
52. D. Montarnal, P. Cordier, C. Soulié-Ziakovic, F. Tournilhac and L. Leibler, *J. Polym. Sci., A: Polym., Chem.*, 2008, **46**, 7925.
53. Y. Chen, A. M. Kushner, G. A. Williams and Z. Guan, *Nature Chem.*, 2012, **4**, 467.
54. L. Fang, M. A. Olson, D. Benitez, E. Tkatchouk, W. A. Goddard III and J. F. Stoddart, *Chem. Soc. Rev.*, 2010, **39**, 17.
55. H. M. Colquhoun, J. F. Stoddart, D. J. Williams, J. B. Wolstenholme and R. Zarzycki, *Angew. Chem. Int. Ed.*, 1981, **20**, 1051.
56. H. M. Colquhoun, E. P. Goodings, J. M. Maud, J. F. Stoddart, D. J. Williams and J. B. Wolstenholme, *J. Chem. Soc., Perkin Trans.*, 1985, **2**, 607.
57. B. W. Greenland, S. Burattini, W. Hayes and H. M. Colquhoun, *Tetrahedron*, 2008, **64**, 8346.

58. Z. Zhu, C. J. Cardin, Y. Gan and H. M. Colquhoun, *Nature Chem.*, 2010, **2**, 653.

59. H. M. Colquhoun, Z. Zhu, C. J. Cardin, Y. Gan and M. G. B. Drew, *J. Am. Chem. Soc.*, 2007, **51**, 16163.

60. H. M. Colquhoun and Z. Zhu, *Angew. Chem., Int. Ed.*, 2004, **43**, 5040.

61. S. Burattini, H. M. Colquhoun, B. W. Greenland and W. Hayes, *Faraday Discuss.*, 2009, **143**, 251.

62. S. Burattini, H. M. Colquhoun, J. D. Fox, D. Friedmann, B. W. Greenland, P. J. F. Harris, W. Hayes, M. E. Mackay and S. J. Rowan, *Chem. Commun.*, 2009, **44**, 6717.

63. S. Burattini, B. W. Greenland, W. Hayes, M. E. Mackay, S. J. Rowan and H. M. Colquhoun, *Chem. Mater.*, 2011, **23**, 6.

64. S. Burattini, B. W. Greenland, D. H. Merino, W. Weng, J. Seppala, H. M. Colquhoun, W. Hayes, M. E. Mackay, I. W. Hamley and S. J. Rowan, *J. Am. Chem. Soc.*, 2010, **132**, 12051.

65. J. D. Fox, J. J. Wei, B. W. Greenland, S. Burattini, W. Hayes, H. M. Colquhoun, M. E. Mackay and S. J. Rowan, *J. Am. Chem. Soc.*, 2012, **134**, 5362.

66. S. J. Eichorn, A. Dufresne, M. Aranguren, N. E. Marcovich, J. R. Capadona, S. J. Rowan, C. Weder, W. Thielemans, M. Roman, S. Renneckar, W. Gindl, S. Veigel, J. Keckes, H. Yano, K. Ade, M. Nogi, A. N. Nakagaito, A. Mangalam, J. Simonsen, A. S. Benight, A. Bismarck, L. A. Berglund and T. Peijis, *J. Mater. Sci.*, 2010, **45**, 1.

67. N. Ouali, J. Y. Cavaille and J. Perez, *Plast. Rubber Compos. Process. Appl.*, 1991, **16**, 55.

68. M. Takayanagi, S. Uemure and S. Minami, *J. Polym. Sci., Part C: Polym. Symp.*, 1964, **5**, 113.

69. J. R. Capadona, O. Ven Den Berg, L. A. Capadona, M. Schroeter, S. J. Rowan, D. J. Tyler and C. Weder, *Nat. Nanotechnol.*, 2007, **2**, 765.

70. M. Burnworth, D. Knapton, S. J. Rowan and C. Weder, *J. Inorg. Organomet. Polym. Mater.*, 2007, **17**, 91.

71. M. Burnworth, L. Tang, J. R. Kumpfer, A. J. Duncan, F. L. Beyer, G. L. Fiore, S. J. Rowan and C. Weder, *Nature*, 2011, **472**, 334.

72. G. L. Fiore, S. J. Rowan and C. Weder, *Chimia*, 2011, **65**, 745.

73. J. R. Kumpfer, J. J. Wie, J. P. Swanson, F. L. Beyer, M. E. Mackay and S. J. Rowan, *Macromolecules*, 2012, **45**, 473.

74. D. Knapton, M. Burnworth, S. J. Rowan and C. Weder, *Angew. Chem. Int. Ed.*, 2006, **45**, 5825.

75. J. R. Kumpfer and S. J. Rowan, *J. Mater. Chem.*, 2010, **20**, 145.

76. J. R. Kumpfer and S. J. Rowan, *J. Am. Chem. Soc.*, 2011, **133**, 12866.

77. K. Kunzelman, M. Kinami, B. R. Crenshaw, J. D. Protasiewiczand and C. Weder, *Adv. Mater.*, 2008, **20**, 119.

78. B. Crenshaw, M. Burnworth, D. Khariwala, P. A. Hiltner, P. T. Mather, R. Simha and C. Weder, *Macromolecules*, 2007, **40**, 2400.

79. A. R. Hurst, B. Escuder, J. F. Miravet and D. K. Smith, *Angew. Chem. Int. Ed.*, 2008, **47**, 8002.

80. M. Suzuki and K. Hanabusa, *Chem. Soc. Rev.*, 2010, **39**, 455.
81. N. M. Sangeetha and U. Maitra, *Chem. Soc. Rev.*, 2005, **34**, 821.
82. M. Nakahata, Y. Takashima, H. Yamaguchi and A. Harada, *Nat. Commun.*, 2011, **2**, 511.
83. A. Phadke, C. Zhang, B. Arman, C-C. Hsu, R. A. Mashelkar, A. K. Lele, M. J. Tauber, G. Arya and S. Varghese, *Proc. Natl. Acad. Sci. U.S.A.*, 2012, **109**, 4383.
84. M. Zhang, D. Xu, X. Yan, J. Chen, S. Dong, B. Zheng and F. Huang, *Angew. Chem. Int. Ed.*, 2012, **51**, 7011.
85. A. Okada and W. Oppermann, *Macromol. Mater. Eng.*, 2006, **291**, 1449.
86. K. Haraguchi and T. Takehisa, *Adv. Mater.*, 2002, **14**, 1120.
87. K. Haraguchi, T. Takehisa and S. Fan, *Macromolecules*, 2002, **35**, 10162.
88. K. Haraguchi, R. Farnworth, A. Ohbayashi and T. Takehisa, *Macromolecules*, 2003, **36**, 5732.
89. K. Haraguchi, H. J. Li, K. Matsuda, T. Takehisa and E. Elliott, *Macromolecules*, 2005, **38**, 3482.
90. Q. Wang, J. L. Mynar, M. Yoshida, E. Lee, M. Lee, K. Okuro, K. Kinbara and T. Aida, *Nature*, 2010, **463**, 339.
91. K. Haraguchi, K. Uyama and H. Tanimoto, *Macromol. Rapid Commun.*, 2011, **32**, 1253.
92. K. Imato, M. Nishihara, T. Kanehara, Y. Amanoto, A. Takahara and H. Otsuka, *Angew. Chem. Int. Ed.*, 2012, **51**, 1138.
93. G. Deng, F. Li, H. Yu, F. Liu, C. Liu, W. Sun, H. Jiang and Y. Chen, *ACS Macro Lett.*, 2012, **1**, 275.
94. P. Sahoo, R. Sankolli, H-Y. Lee, S. R. Raghavan and P. Dastidar, *Chem. Eur. J.*, 2012, **18**, 8057.
95. G. R. Desiraju, *Angew. Chem. Int. Ed.*, 1995, **34**, 2311.

CHAPTER 5

Thermodynamics of Self-Healing in Polymeric Materials

YING YANG AND MAREK W. URBAN*

Department of Materials Science and Engineering, Clemson University, Clemson, SC 29634
*Email: mareku@clemson.edu

5.1 Introduction

The target of producing man-made materials in general, and polymers in particular, capable of repeatedly repairing themselves is driven by a desire to create sustainable and environmentally compliant technologies.[1-3] Often inspired by biological systems, the self-healing of polymers is being viewed as perhaps one of the most attractive and technologically advanced features of future materials, particularly when repeatability and reversibility of the damage–self-healing cycle can be achieved. The fundamental difference between biological and synthetic materials is the ability of the former to metabolize; that is, to effectively utilize or remove side products produced during self-healing cycles. Since synthetic materials do not exhibit these attributes, design of efficient self-healing polymers is challenging and requires an interplay of physico-chemical processes, which may or may not be reversible, at both the molecular and macroscopic scale. Intuitively, self-healing involves conformational changes of macromolecular segments in parallel with bond reforming chemical reactions, leading to network remodeling. The complexity of these transient events is compounded by numerous covalent and non-covalent bond reformations, which may or may not be reversible, often in parallel with conformational and configurational changes of macromolecular

RSC Polymer Chemistry Series No. 5
Healable Polymer Systems
Edited by Wayne Hayes and Barnaby W Greenland
© The Royal Society of Chemistry 2013
Published by the Royal Society of Chemistry, www.rsc.org

segments which possess different flexibilities. Thus thermodynamic states of macromolecular chains and chemical reformability of individual segments will determine the success of self-healing reactions. The first part of this chapter focuses on thermodynamic aspects of self-healing and the role of the entropic and enthalpic components of the Gibbs free energy in the context of recoupling the self-healing lattice model that accounts for the reactivity and flexibility of macromolecular segments. The second section of this chapter outlines recent notable advances in chemistries utilized in self-healing of thermoplastic and thermosetting polymers.

5.2 Thermodynamics of Self-Healing

Initial studies on thermoplastic polymers suggested that crack healing involved five stages: segmental surface rearrangements, surface approach, wetting, diffusion, and randomization.[4] In essence, the outcome of this process was to regain mechanical properties at the repaired area through forming new chain entanglements. Thus chain diffusion at the polymer–polymer interface was considered the primary driving force for repairs, and the reptation model[5] semi-quantified these events by describing the motion of macromolecular segments trapped in a stationary tube. Under these boundary conditions, self-healing was achieved only when the entire chain completely disengaged from the tube, reaching an equilibrium state after time T_r (the time required for complete chain disengagement from the tube). Using this concept, the motion of one chain end can be characterized by the relaxation of the chain end by using Gaussian partition function Z, Equation (1),[6] where ξ is the distance of an end segment from the interface, ξ_1 and ξ_2 represent the positions of the two ends of one chain, $\gamma = 1/2a^2N$, a is the length of one segment, and N is the number of segments of a chain.

$$Z(\xi_1, \xi_2) = (\gamma/\pi)^{1/2}[exp\{-\gamma(\xi_1 + \xi_2)^2\} + exp\{-\gamma(\xi_1 - \xi_2)^2\}\}] \qquad (1)$$

When the equilibrium conformation is achieved, chains become Gaussian in nature, and Z can be expressed as shown in Equation (2).

$$Z_e(\xi_1, \xi_2) = (\gamma/\pi)^{1/2}exp\{-\gamma(\xi_1 - \xi_2)^2\} \qquad (2)$$

Under these conditions, the self-repairing process can be viewed as chain conformational changes from non-Gaussian to equilibrated Gaussian state, during which each chain escapes from the tube in a given time t, and its length is given by Equation (3)

$$\langle l \rangle \approx 2(Dt/p)^{1/2}[1 - exp\{-L^2/2Dt\}] \qquad (3)$$

The term $\langle l \rangle \propto t^{1/2}$ represents the early stages of self-healing; for short times, the exponential term is small, but to complete the repair, each chain must diffuse out of the tube. Considering molecular weight dependence,[4] the repair time follows $T_r \propto M^3$ relationship, suggesting that lower molecular weight polymers exhibit favorable repairing conditions. Assuming bond cleavage does not occur during damage, chain diffusion is the only driving force for repair, and the source of loose chain ends is chain slippage. The healing efficiency defined by the extent of the recovery with respect to its initial post-damage state can be expressed as the ratio of stress, $R(\sigma)$, before and after healing (4).

$$R(\sigma) = \frac{\sigma_{healed}}{\sigma_{initial}} \quad (4)$$

It has been also shown that the fracture stress σ is proportional to $(t/M)^{1/4}$ $(\sigma \propto (t/M)^{1/4}$; where M is polymer molecular weight).[6] Mathematically, if repair is considered as the rearrangement of chain conformations from non-Gaussian to Gaussian states, the time dependent evolution of chain lengths, as well as the fracture stresses, will again favor low molecular weight polymers that exhibit shorter repairing times and energetically favorable stress recoveries. These chains also exhibit lower glass transition temperatures (T_g), in particular near the interfacial regions. However, chain slippage will not be the sole source of active chains. During repair, physical or chemical damage often will cause chain ruptures, resulting in reactive chain ends that significantly contribute to self-healing, as will be discussed later.

While prior studies are useful when diffusion is the primary driving force for self-healing, self-repairing involves other physico-chemical interactions at various timescales. Recent studies have shown that favorable molecular inter-actions within a polymer matrix will facilitate self-healing *via* supramolecular chemistry or multi-level covalent bond reformation. As damage occurs, chains cleave at the weakest points, such as hydrogen bonds, metal-ligand coordination sites, or weakest covalent bonds. Although covalent bonds may require higher energies to cleave, they typically generate free radicals when cleaved, the stability of which is critical for achieving desirable repairs. Regardless of the energy input and chemistries involved, mobile and reactive chain ends capable of bond reforming will exhibit higher kinetic and potential energies. At the same time, release of constraints in localized areas of the surrounding polymer matrix will be manifested by relaxations of loose chain entanglements at newly created interfaces. This is schematically depicted in Figure 5.1A.

After damage, being a non-equilibrium state, self-healing will result in a new Gaussian equilibrium state *via* localized reactions and associated chain relaxations, primarily at loose chain ends. Energetically, self-repair will occur when the Gibbs free energy changes $(\Delta G = \Delta H - T\Delta S)$ are <0. In the early stages, entropic contributions are the primary driving force; as soon as the chain ends react and approach the Gaussian state, the enthalpic term will dominate ΔG. This is schematically shown in Figure 5.1B.

The following section will consider the role of enthalpic and entropic contributions, showing that they are significantly different at different stages of self-healing of thermosetting and thermoplastic polymers, and that entropic contributions play a dominant role for higher disordered networks.

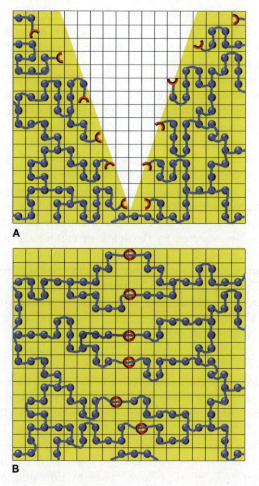

Figure 5.1 Recoupling self-healing lattice model of free chain ends at damaged polymer interface. A represents the damaged state with dangling chains and voids; B is the repaired state with free chain ends cross-linked with their counterparts. Each square represents a single lattice site occupied by one segment of a polymer chain. Open circles represent reactive chain ends, whereas full circles are remaining segments of the lattice.

Assuming that all the chains are energetically equal and there is no energy exchange between neighboring chains, the entropy change for a finite change from state 1 (Figure 5.1A) to state 2 (Figure 5.1B) during repair can be expressed as Equation (5), where k is the Boltzmann constant, and Ω_1 and Ω_2 are the numbers of available arrangements of chain segments at state 1 and state 2.

$$\Delta S = S_2 - S_1 = k \ln(\Omega_2/\Omega_1) \quad (5)$$

Analogous to spontaneous expansion of gas molecules in an open space, the volume increase near created interfaces will lead to an increase in the number of arrangements available for a given state. Therefore, driven by the positive entropy changes, chain ends will continuously expand into voids, particularly when two surfaces are in favorable distances. However, unlike freely diffusing small molecules, polymeric chains exhibit limited mobility within the confinements of a polymer matrix.

In view of these considerations, it can be assumed that there are two types of macromolecular domains represented as self-healing recoupling lattices, generated as a result of mechanical damage: (1) free chains resulting from chain rupture with both chain ends being free, and (2) tethered chains having one end anchored to the surface and the other free end dangling, as depicted in Figure 5.1. Assuming that free chains exhibit both translational and configurational freedom, the latter aspect of tethered chains will be the primary contributor to the entropy changes. Furthermore, the confinements of chains near the anchored surface will result in smaller entropy changes, which will tend to increase toward their dangling ends. Since entropy changes, as well as self-repair attributes, are affected by flexibility of chains; the influence of chain rigidity, reflected by the glass transition temperature, for a given polymer will have a significant effect on the number of chain configurations, as defined in Equation (1). Depicted in Figure 5.1A and B, is a self-healing recoupling lattice model illustrating two states, state 1 and state 2, immediately after damage (Figure 5.1A) and during repair (Figure 5.1B), respectively. As shown in Figure 5.1, for n self-avoiding walk polymer chains, each has N segments (a segment is defined as a repeating unit or a macromolecular unit that maintains its conformation during repair), $(N-1)$ bonds, and $(N-2)$ bond angles, fitted into a self-healing recoupling lattice, where each segment occupies one lattice site. The entropy changes will be principally localized near the damaged area, which is primarily occupied by dangling ends (half circles, Figure 5.1). If one considers chains being semi-flexible, the flexibility parameter, f, is defined as the probability of two successive bonds not being collinear and are subject to rotations, so that out of $n(N-2)$ bonds, there will be $fn(N-2)$ bent bonds. Under the condition that $\partial\triangle G/\partial f = 0$, f can be given by Equation (6),[7] where z is the coordination number of the lattice unit, k_B is Boltzmann constant, T is the absolute temperature, and ε is the energy difference between collinear and bent bonds.

$$f = \frac{(z-2)\,e^{-\varepsilon/k_B T}}{1 + (z-2)e^{-\varepsilon/k_B T}} \tag{6}$$

The primary consequence of mechanical perturbations of the lattice will be lowering molecular weight and removal of a fraction of polymer lattice sites, which will parallel the formation of new sites. The total number of chain conformations at the damaged site can be expressed by Equation (7),[7] where ν_j accounts for the total number of chain end configurations.

$$\Omega = \frac{1}{n!2^n} \left(\frac{(N-2)n}{f(N-2)n} \right) \prod_{j=1}^{n} \nu_j \tag{7}$$

Since Equation (7) provides the binomial distribution of chain end configurations that account for the probability of finding $fn(N-2)$ bent bonds out of a total of $n(N-2)$ bonds (assuming that the energy of molecular interactions is zero), ΔG for free chains with two ends free can be given by Equation (8).[7]

$$\Delta G = -T\Delta S = -nRT\{\ln N + \ln(z/2e) + (N-2)\ln[1/(1-f)e]\} \qquad (8)$$

The first term, $\ln N$, originates from the opportunity for the first chain segment to be in a random location. For tethered chains, assuming that the first segment remains in a fixed position during repair, $N=1$ and $\ln N=0$ and the remaining terms of Equation (8) will apply (assuming that ε is the same for all bonds, although ε may be larger at the tether end than at the free chain end). Thus ΔG for tethered chains becomes Equation (9).

$$\Delta G = -T\Delta S = -nRT\{\ln(z/2e) + (N-2)\ln[1/(1-f)e]\} \qquad (9)$$

According to Equations (8) and (9), f will have the threshold value K, such that when $f>K$, $\Delta G<0$ and the chains will be able to move spontaneously. When $f<K$, the chains will be too rigid to move, making repair unlikely to occur, or requiring higher external energy inputs.

For free chains, it can be shown from Equation (8) that:

$$K_F = 1 - (1/e)(zN/2e)^{1/N-2} \qquad (10)$$

Whereas for tethered chains, K becomes defined by Equation (11).

$$K_T = 1 - (1/e)(z/2e)^{1/N-2} \qquad (11)$$

Assuming that tethered chains consisting of 10 segments have the co-ordination number $z=6$, for self-healing to occur, it is required that $K_T=0.63$ and $f>0.63$. For a free chain with 10 segments, however, $K_F=0.52$, implying that to initiate conformational changes, lower flexibility will be necessary. Since the K value increases with increasing number of chain segments, therefore longer chains will require higher flexibility to acquire movement; but the difference for the chains with 1000 and 10 segments can be as low as 0.01.

According to Equation (6), at $T=$constant, f will be determined by the energy difference between bent and collinear bond states, ε. As the self-repairing process depicted by the recoupling self-healing lattice model in Figure 5.1 continues, chain ends come into contact with each other to form new cross-linked sites or entanglements, leading to the recovery of mechanical properties.

Free chains exhibit higher f values as a result of lower ε, and typically have smaller N values compared to the tethered chain population, which results in more negative ΔG and greater mobility. When free chains react with each other and add onto tethered chains, they contribute to the increase in N. As this process continues, chains become more rigid and, as N increases until f reaches the threshold value of K for a given N, ΔG will be positive. At this point, a stable Gaussian equilibrium state is reached. When $f<K$ after damage, repair will not occur. According to 9, low molecular weight, as well as low T_g (low ε, high f), polymers will favorably contribute to higher ΔS values, resulting in $\Delta G<0$. These attributes will facilitate chain end motions toward repair, and are one of the prerequisites, but not the sole requirement, for self-healing.

Self-repair *via* diffusion may not be achieved when enthalpic components dominate macromolecular miscibility unless inter and intramolecular interactions defined by the Flory-Huggins parameter, χ, are favorable, or an external energy source is applied.

Considering internal energy contributions to self-repair, the internal energy change is defined by: $\Delta U = \Delta Q - W$, where U is the internal energy, ΔQ is the heat, and W is the work done to the surroundings or any other form of external energy input. Since enthalpy is defined as $H = U + pV$ (p – pressure, V – volume), at p = constant, the internal energy $\Delta U = \Delta Q - P\Delta V + E_{ext}$, (where: E_{ext} is any other form of external energy input to the system except Q and W). This may be expressed as Equation (12),

$$\Delta H = \Delta Q + E_{ext} \qquad (12)$$

assuming that p = constant during repair, ΔH is the sum of heat transfer, Q, associated with chain rearrangements and chain end reactions at the interface, and any other forms of energy input from damaged and repaired states. In addition to entanglements of chain ends at two surfaces, reactive counterparts at each chain end (for example non-hydrogen bonded ends), non-coordinated sites, or reactive free radicals, will contribute to ΔH. These reactive macromolecular segments generated by damage should retain their reactivity in order to participate in self-healing reactions. Thus interactions with neighbors from the same interface (Figure 5.1A), or quenched by any other gas phase molecules, will not favor self-healing reactions.

Energetic contributions to ΔH are critical and, at the onset of repair, may need to be triggered by built-in stimuli-responsive components. For example, external stimuli can be utilized to break certain linkages, which may enhance mobility or create reactive sites for future reactions. Ideally, these stimuli network components need to be able to reform after repair, otherwise repair may not be reversible. Scheme 5.1 illustrates the reaction of two macromolecular segments that contain stimuli-responsive components (STRs) and reactive chain ends X and Y. A fraction of STR that is unable to recover to its original state is termed ST, and E (Scheme 5.1) represents all stimuli energy input including E_{ext}, Q and W. The enthalpy change associated with the healing process can be approximated by bond energies using Equation (13), where ω is the sum of the bond energies of each component.

$$\Delta H = [\omega_{X-Y} - (\omega_X + \omega_Y)] + [\omega_{ST} - \omega_{STR}] \qquad (13)$$

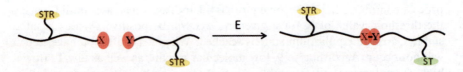

Scheme 5.1   A schematic illustration of reactions between two free chain ends (X and Y) and the changes of stimuli-responsive components (STR).

If the chain end groups react with neighbors from the same surface or are quenched by other species, the total enthalpy changes may be similar, but these conditions are undesirable for self-healing.

Damage–repair processes can be reversible or irreversible. When stimuli-responsive component STR is present, it absorbs energy in order to be activated to a higher energy state and trigger the onset of repairing. After removal of the stimuli, a fraction of STR may recover, giving off heat. If STR is completely recovered, the process shown in Scheme 5.1 can be carried out again and be "reversible self-repairing". If STR cannot be recovered at all, only one-time repair, "irreversible repair", is anticipated. STR may also be partially recovered, and a limited number of repair cycles are expected.

For irreversible, infinitesimal irreversible self-healing processes to occur, $\Delta S > \Delta Q/T$. Substituting ΔQ with Equation (12), $E_{ext} > \Delta H - T\Delta S$, and repair process will occur according to Equation (14).

$$\Delta G - E_{ext} < 0 \qquad (14)$$

The most efficient self-healing can be achieved when chain ends recouple exothermically ($\Delta H < 0$), with $\Delta S > 0$. One example of such networks is polymers with the ability of free radical recoupling or hydrogen bonding formation. At the same time, if chain mobility is high enough to facilitate diffusion, external energy input may not be necessary. When self-healing reactions are endothermic ($\Delta H > 0$), or a polymer network is too rigid to undergo conformational changes ($\Delta S < 0$), E_{ext} will be required to overcome the positive ΔG. As indicated earlier, E_{ext} may activate built-in stimuli-responsive components, leading to increased kinetic energy of the chains, thus decreasing the energy barrier for chain end recoupling reactions.

Although each polymer system will have its own characteristics, in general—for polymeric solids, especially polymer gels—entropic contributions favor self-repairs. As chains undergo elastic deformations and inter-diffusion at localized areas, they will take new conformations and ultimately fill up the voids created by mechanical damage. If bond formation during repair is exothermic, it contributes to more negative ΔG values. However, if the activation of stimuli-responsive components (STR) requires external energy input (E), or reactive chain ends can be quenched to another ground state before repair, additional external energy input will be needed to overcome the new energy barriers. Realizing that each self-healing process involves chemical reactions and physical network rearrangements, the nature of self-healing will be specific to a given polymer matrix. Figures 5.2A to C illustrate the general energetic contributions of stimuli-responsive behavior for solutions, gels, and solids.[8]

The transitions between these minima represent the energy required for a system to go from one state to another, which can be easily achieved in solutions when Brownian motion energies are low and conformational changes require small energy inputs. This is illustrated by two energy minima, shown in Figure 5.2A, which facilitate the equally easily achievable energy states in two different conformations. For gel networks with higher density, yet typically low

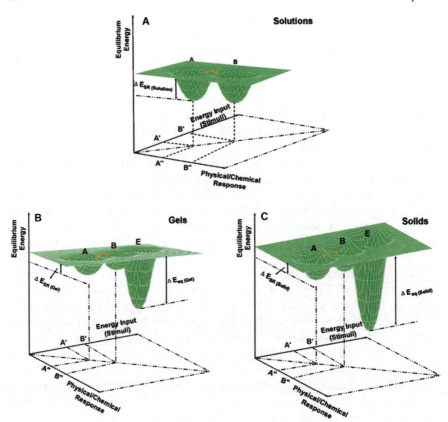

Figure 5.2 The relationship between equilibrium energy, stimuli energy input, and physical/chemical response for solutions (A); gels (B); and solids (C). Reproduced with permission from reference [8].

T_g, one needs to take into account the stability of the network, illustrated by the lower energy state, $\Delta E_{eq\ (Gel)}$, shown in Figure 5.2. This is associated with two smaller energy minima, A and B, representing the stimuli-responsive transition between STR and ST, a prerequisite for self-healing. When considering the solid state, the primary difference is the depth of the minimum energy of $E_{eq(Solid)}$ responsible for network stability. For network repairing, A (STR state) and B (ST state) will provide stimuli-responsiveness responsible for self-healing.

The main question is how these energy distributions, defined by chemical makeup of the network, can be utilized in self-healing process. When damage occurs, the lowest energy state, $E_{eq(Solid)}$, will rapidly change to a less stable state and macromolecular segments will have high mobility, resulting in lower T_g values, near the surface/interface created by damage. At the same time, energy contributions from self-healing elements will exhibit higher energy states in order to respond to damage and initiate the repair process. Upon repair, the $E_{eq(Solid)}$ of solids will be restored, but A and B may or may not be restored.

If self-healing is reversible, A and B will regenerate again to the lower energy levels, at the expense of new equilibrium energy; but for irreversible self-healing, A minimum (STR state) will vanish.

While these general considerations may serve as a guide for designing self-healable thermoplastic and thermosetting polymers, it should be realized that precise molecular design with controllable polymer topologies are necessary to meet thermodynamic requirements. The last decade has offered a variety of chemical approaches to self-healing in thermoplastic and thermosetting polymers, of which some are reversible, and some are not. The following section outlines recent advances in thermoplastic and thermosetting polymers that exhibit both reversible and reversible characteristics.

5.3 Self-Healing Reactions in Thermoplastic and Thermosetting Polymers

Regardless of the thermoplastic or thermosetting nature of a given polymer, the recoupling self-healing lattice model provides an estimate of which of the network components predominantly contributes to the transition between non-equilibrium and states during self-repair. Parallel to conformational and configurational changes, chemical events that contribute to self-healing are:

(1) bond cleavage, which will generate reactive bond-reformable groups, such as COOH, NH_2, OH, SH, C=O, or free radicals;
(2) chain slippage, which will result in configurational and conformational changes;
(3) combination of supramolecular networks with reforming hydrogen bonding or metal-ligand coordination bonds.

5.3.1 Bond Cleavage Reactions in Thermoplastic and Thermosetting Polymers

Since free radicals or other reactive chain ends generated by cleavage reactions may be intercepted by oxidative or other processes, one of the key challenges is facilitating stability of reactive groups during the network repair. For example, regardless of molecular weight, polycarbonate (PC) cannot self-repair upon mechanical damage. However, synthesis of PC capable of self-repairing by ester exchange reactions can be accomplished using a steam pressure at 120 °C.[9] As shown in Figure 5.3A, hydrolysis of carbonate groups under steam pressure generates phenolic chain ends, which can undergo ester exchange catalyzed by $NaHCO_3$ to recombine chain ends. As illustrated in Figure 5.3B, the repair process for poly(phenylene-ether)s (PPEs) will work in a similar manner.[10–12]

Damages along the PPE backbone produce stabilized oxygen radicals by their complexation with Cu(II), which further inhibits oxidation, and the presence of hydrogen donors facilitates their conversion to phenol groups. Thus phenolic chain ends are able to recombine through ether exchange

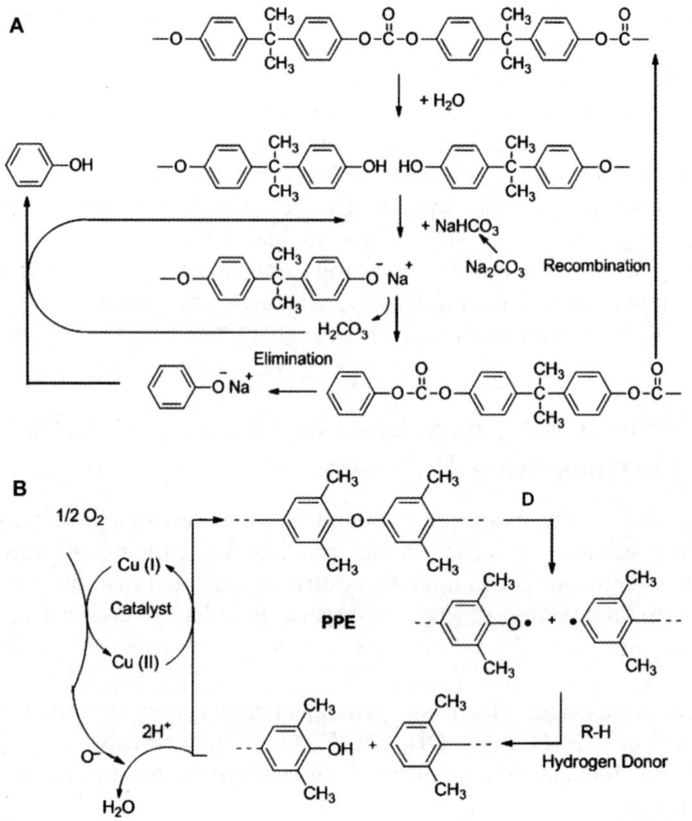

Figure 5.3 (A) ester exchange reaction in polycarbonate leading to healing;
(B) healing in polypropylene ethers *via* covalent bond formation between
scission chain ends in the presence of a Cu catalyst.
Adapted from Reference [9 and 12].

reactions at elevated temperatures, catalyzed by Cu(II). These examples
illustrate that built-in stimuli-responsive components (which are carbonate
groups sensitive to pressurized steam) and hydrogen donors, respectively,
enable recombination of chain ends leading to self-healing. It should be also be
noted that PC and PPE are both high molecular weight rigid polymers, and
ester and ether exchange reactions are endothermic in nature. Therefore high
temperatures are required to overcome the energy barriers of chain flexibility
and elimination reactions in the presence of catalysts. These conditions meet
energetic criteria related to ΔG and E_{ext} relations discussed previously [Equation
(14)]. Furthermore, chain ends of these thermoplastic macromolecules are
stabilized at the expense of consumption of stimuli-responsive reactants,
and bond reforming reactions responsible for self-healing require elevated
catalyst levels. Also, if concentration levels of stimuli-responsive components are
excessive, mechanical properties may be altered.

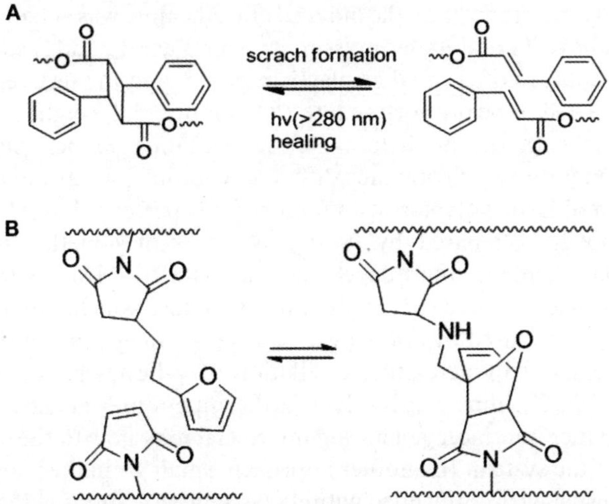

Figure 5.4 (A) [2 + 2] cycloaddition reactions of cinnamoyl groups leading to healing; (B) Diels–Alder reactions of multi-furan and multi-malemide containing backbone within cross-link polymer system for reversible self-repairing. Adapted from references [13] and [20].

To minimize the influence of added components, cross-linked networks with reversible bonds were utilized to achieve stable yet reactive chain end entities after damage, as well as to achieve covalent bond-reformation, enabling network repair. External energy input in the form of UV radiation and heating will activate the reversible reactions without chemical input, thus minimizing by-product formation. Figure 5.4A illustrates reactions of cinnamate monomer 1,1,1-tris-(cinnamoyloxymethyl)ethane (TCE) cross-linked *via* a [2 + 2] photocycloaddition, resulting in high T_g transparent films.[13] Mechanical damage leads to the C–C bond cleavage of cyclobutane rings between TCE monomers, resulting in the formation of original cinnamoyl groups, whereas the crack-healing characterized by recovery of flexural strength occurs by re-photocycloaddition to reform cyclobutane ring cross-links upon UV (>280 nm) exposure at 100 °C.[14–17]

Another approach illustrated in Figure 5.4B utilized thermal reversibility of the Diels–Alder (DA) and retro-DA reactions in furan–maleimide, dicyclo-pentadiene, and anthracene based polymers.[18–21] These reversible reactions were applied to epoxy, acrylate, and polyamide systems, where retro-DA reactions resulted in disconnection of butylene oxide ring, leading to the formation of diene and dienophile. The reconstruction of covalent bonds, repairing the crack, occurs *via* recoupling of diene and dienophile groups into butylene oxide rings. Again, reactive chain ends capable of reforming linkages during repair are critical to the successful design of self-healing polymers.

As was demonstrated in Section 5.2, interfacial diffusion is one of the contributing elements to favorable Gibbs free energy changes during self-healing, and is enhanced in a melted state for polymers with high flexibilities, f_F.

In poly(methyl methacrylate), the induced crack healing was achieved by heating the sample above T_g and under pressure.[22] To reduce T_g and facilitate healing, organic solvents are often used as plasticizers,[23,24] and the recovery of fracture stress is suggested to be proportional to $t^{1/4}$ (if molecular weight M is constant), thus agreeing with the predictions of the reptation model. Since localized elevated temperatures will enhance diffusion without solvent interference, self-healing thermoplastic polymers with uniformly dispersed γ-Fe_2O_3 nanoparticles were developed.[25] Prepared by *in situ* synthesis of *p*-methyl methacrylate/ *n*-butylacrylate/heptadecaflurodecyl methacrylate (*p*-MMA/*n*BA/HDFMA), colloidal particles coalesce to form uniform films, which, upon mechanical damage, can be repaired by applying an oscillating magnetic field (OMF). Self-healing is achieved by activating oscillation of γ-Fe_2O_3 nanoparticles at the frequency of OMF, thus causing localized temperature increase at the nano-particle–polymer interface, generating interfacial melt flow to the damaged area and repair of the system. In another approach, small grain thermoplastic epoxy particle adhesives were embedded within glass-epoxy composite matrices, which melted upon heating, thus repairing damages.[26] Although the voids left after melting of adhesives may have adverse effects on the integrity of the epoxy polymer matrix, this approach may serve many applications. Along the same lines, blending of thermoplastic poly(bisphenol-A-co-epichlorohydrin) with epoxy resin at 80 °C to dissolve it into the thermoset was utilized.[27] During damage, thermoplastic polymer in the vicinity of the crack flows to the newly created crack regions and heals the crack.

Ionic interactions in polymer matrices have been known to enhance mechanical properties, and selected polymers have shown to exhibit self-healing attributes. In poly(ethylene-co-methacrylic acid) and polyethylene-*g*-poly(hexylmethacrylate),[28] instead of external heat or an alternative stimulus (E_{ext}) to promote polymer diffusion in a damaged area, repairs were facilitated under ambient and elevated temperatures upon projectile puncture testing.[29,30] A ballistic puncture in low-density poly ethylene does not exhibit self-healing itself, whereas puncture in poly(ethylene-*co*-methacrylic acid) films heals, leaving a scar on the surface. The proposed healing mechanism is a two stage process in which, upon projectile impact, the ionomeric network is disrupted and heat generated by the friction during the damage is transferred to the polymer surroundings, generating localized melt state. Elastic responses of the locally molten polymer facilitated by a puncture are known as puncture reversal. During the second stage, the molten polymer surfaces fuse together *via* interdiffusion, resulting from ionic interactions between ionic clusters, to seal the puncture, followed by rearrangement of the clustered regions and long-term relaxation processes, which continue until ambient temperatures are reached.

Combining repair and sensing attributes in one polymer matrix was accomplished in the colloidal synthesis of poly(methyl methacrylate/ *n*-butylacrylate/2-[(1,3,3-trimethyl-1,3-dihydrospiro[indole-2,3′-naphtho[2,1-b]-[1,4]oxazin]-5-yl)amino]ethyl-2-methylacrylate) (Figure 5.5A). These films, upon mechanical damage, undergo a change from colorless to red in damaged areas. However, upon exposure to sunlight, temperature and/or acidic vapors,

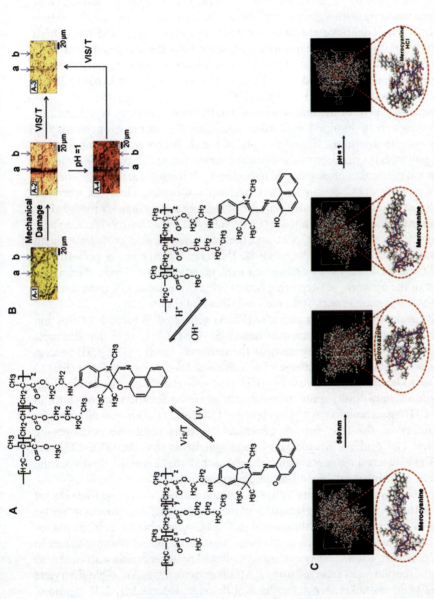

Figure 5.5 (A) Stimuli-responsiveness of self-healing copolymer, poly(methyl methacrylate/n-butylacrylate/2-[(1,3,3-trimethyl-1,3-dihydrospiro[indole-2,3'-naphtho[2,1-b][1,4]oxazin]-5-yl)amino]ethyl-2-methylacrylate) [p(MMA/nBA/SNO)]; (B) optical images show color change upon scratch and self-healing behavior; (C) chain conformational changes facilitated by SNO ring-opening and closure.
Adapted from reference [31].

damaged areas are repaired and the initial colorless appearance is recovered. As shown in Figure 5.5B, the process is reversible and driven by conformational changes of the statistical copolymer that parallels the ring opening–closure of spironapthoxazine (SNO) groups.[31] Upon mechanical damage, SNO segments are converted into ring opened state merocyanine (MC) forming intermolecular hydrogen bonding with neighboring segments, thus stabilizing the copolymer backbone. The backbone reaches an extended conformation and has a high mobility; this facilitates expansion of polymer into the damaged area. The following ring-closure of MC to form SNO, resulting conformational change of the polymer chains to a coiled state, leads to chain entanglements that facilitate the repair. This is illustrated in Figure 5.5C.

These examples illustrate that self-healing in thermoplastics can be achieved by the introduction of localized molecular heterogeneities in the form of reactive groups or nano-insertions. In contrast, one of the challenges in designing thermo-setting polymers is to incorporate self-healing components while maintaining useful mechanical properties, often reflected by higher T_g. Polyurethanes (PUR) represent these types of networks, but are unable to undergo self-healing. However, when one of the most abundantly available natural polysaccharides, chitin, is modified by acetylation reactions to form chitosan (CHI), followed by further modification with the four-membered oxetane (OXE), and the macromonomer is cross-linked with tri-functional hexamethylene di-isocyanate (HDI) in the presence of polyethylene glycol, these PUR networks exhibit self-healing properties.[32] Upon mechanical damage of the network, self-repairing occurs upon exposure to UV radiation. The reactions leading to network formation are shown in Figure 5.6.

PUR, in combination with polyurea (PUA) generated as a result of reaction between isocyanate and amine-functionalized OXE-CHI, provides desirable integrity and localized heterogeneity of the network, but the OXE-CHI macro-monomer facilitates both cleavage of a constrained 4-membered ring (OXE) to form free radicals, and UV sensitivity (CHI) for self-repair.[33] Similar self-healing, although mechanistically quite different, was achieved for the oxolene-chitosan (OXO-CHI) macromonomer.[34] The choice of OXE and OXO rings was driven by the stability of the free radicals generated by their respective ring-opening processes. The primary advantage of these systems is that the OXE-CHI and OXO-CHI macromonomers can be co-reacted to existing thermosetting polymers, ensuring desirable self-healing characteristics.

Dynamic covalent chemistry offers a unique alternative for self-healing of thermosetting polymers and gels capable of maintaining their mechanical integrity. As shown in Figure 5.7A, trithiocarbonates (TTCs) are able to undergo photo-stimulated reshuffling reactions in the same manner as that of photoinitiators in RAFT polymerization.[35–37] A covalent cross-linked polymer matrix with enhanced segmental mobility was achieved using RAFT copolymerization of n-butyl acrylate and a TTC cross-linker. Upon mechanical damage, followed by UV exposure, repetitive network repairs were achieved, making this network a unique combi-nation of dynamic covalent bond reshuffling reactions and network remodeling.

Disulfide (S–S) bonds, which can be reversibly reduced to thiol (S–H) groups (Figure 5.7B), offer a self-healing alternative.[38] Poly(n-butyl acrylate) grafted

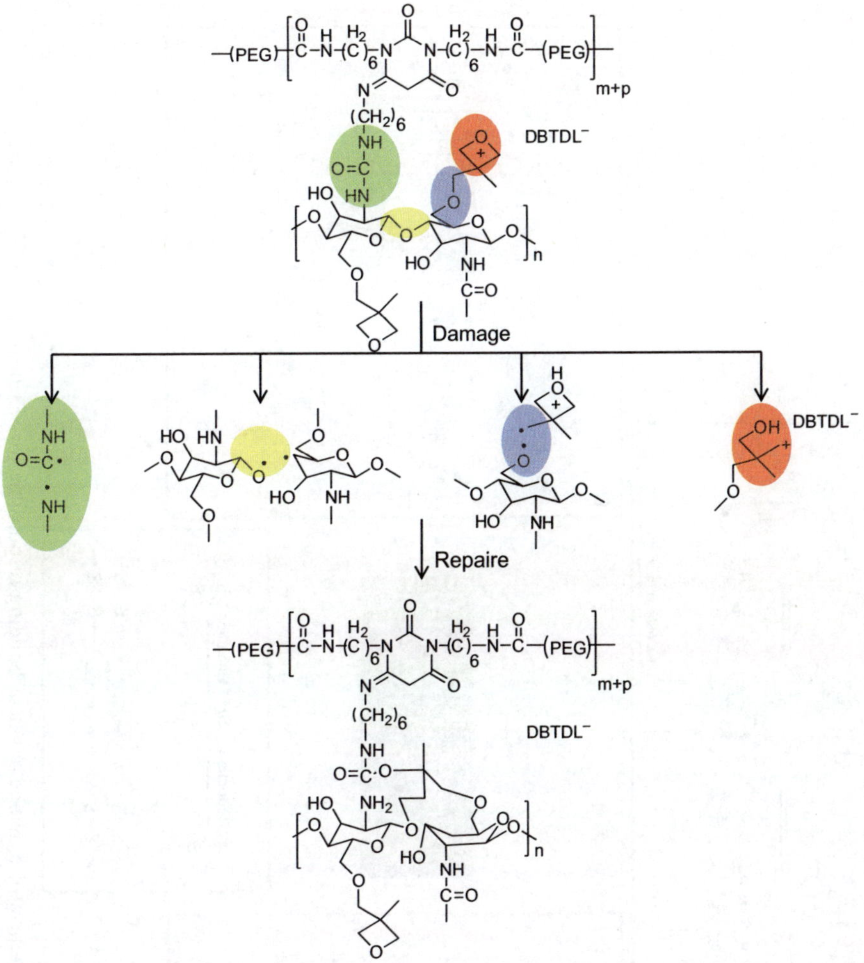

Figure 5.6 Synthetic steps involved in creating self-healing polyurethanes (PURs) containing oxetane (OXE) and chitosan (CHI). Adapted from reference [33].

star polymers were prepared by chain extension atom transfer radical polymerization from cross-linked cores comprised of poly(ethylene glycol) diacrylate) and further utilized as macroinitiators for the consecutive chain extension of bis(2-methacryloyloxyethyl disulfide).[39] This approach introduces reversible disulfide cross-links into the branch peripheries of cross-linked networks that, upon cleavage under reducing conditions, form thiol-functionalized, soluble star polymers. Grafted polymer architectures facilitate intrinsically low viscosity and high accessibility of functional groups, critical to self-healing. Disulfide exchange reactions were also incorporated into low T_g gel networks to achieve room temperature reversibility.[40]

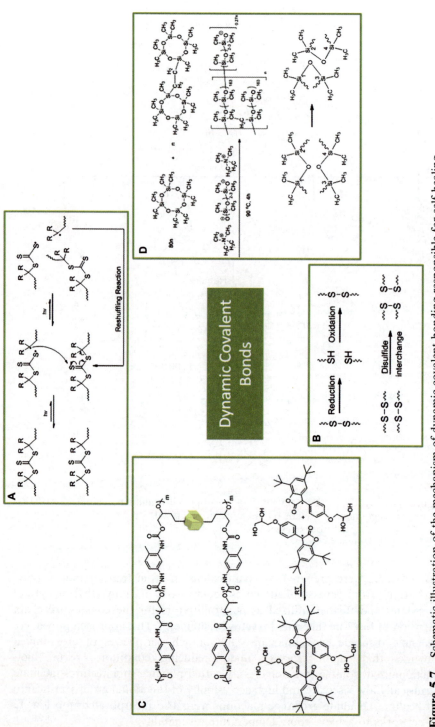

Figure 5.7 Schematic illustration of the mechanism of dynamic covalent bonding responsible for self-healing. Adapted from references [36], [38], [40], [41], and [46].

Figure 5.7C illustrates the formation of diarylbibenzofuranone (DABBF) dynamic covalent bonds *via* stable arylbenzofuranone (ABF) radicals, enabling reformation of DABBF at room temperature, thus facilitating self-healing.[41] These self-healable thermosetting hydrogel networks were prepared by reacting tetrahydroxy-functionalized DABBF building blocks with a modified poly(propylene glycol) bearing an isocyanate group at each end. As shown in Figure 5.7C, when DABBF dissociates to give pairs of ABF radicals, it is critical that the radicals exhibit little or no sensitivity to oxygen,[42] providing sufficient time for covalent bonds to reform during network remodeling,[1] otherwise, oxidative processes will intercept the process.

Silicone-based polymers exhibit a broad spectrum of properties and can be either thermosets or thermoplastics. They are known for their ability to dynamically restructure themselves under various conditions,[43,44] and the recently revisited tetramethylammoniumsilanolate-initiated ring-opening copolymerization of octamethylcyclotetrasiloxane (D4) and bis(hepta-methylcyclo-tetrasiloxanyl)ethane (bis-D4) showed again that these cross-linked polymers containing ethylene bridges and active silanolate end groups exhibit remodeling attributes.[45] Figure 5.7D depicts reactions leading to the formation of "living" tetramethylammoniumdimethylsilamoate end groups, which maintain their activity under ambient conditions, thus providing "ready-to-respond" active groups.

5.3.2 Supramolecular Approaches to Healable Thermoplastic and Thermosetting Polymers

Numerous studies have shown how supramolecular chemistries may facilitate bond reforming upon exposure to physical stimuli. Figure 5.8A illustrates how combining four hydrogen bonds in a functional unit of ureaisopyrimidone (Upy) leads to enhanced association strengths between Upy units and poly-siloxane, polyethers, and polyesters.[46–48] Although relatively weak, hydrogen bonding is the key bond reforming ingredient, and, as a result of their highly directional nature, multiple hydrogen bonds are able to facilitate sufficient mechanical strength to form thermoplastic elastomers.[49] Based on this concept, a self-healing rubber was synthesized in two steps using fatty diacids and triacids from renewable resources.[50] As shown in Figure 5.8B, the first step involves condensation of acid groups with an excess of diethylenetriamine; the second step involves a reaction with urea to obtain a non-tacky rubber-like plastic with $T_g = 8\,^\circ C$, by adding dodecane as plasticizer. Multiple hydrogen bonds are formed between N–H and C=O groups within the network, while crystallization is avoided. As a result of the advantage of high segmental mobility in hydrogels, other soft materials capable of self-healing utilizing hydrogen bonding with temperature or pH dependence have been developed.[51]

Copolymers provide the opportunity for designing multi-phase hetero-geneous materials offering a broad range of mechanical properties. Figure 5.8C

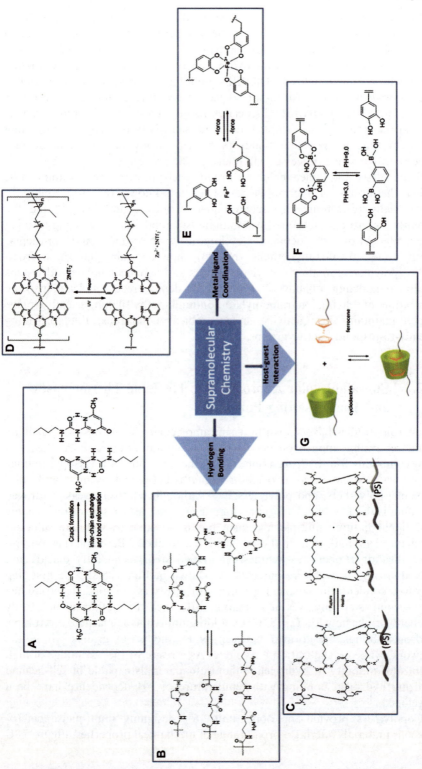

Figure 5.8 Self-healing model of supramolecular interactions, including hydrogen bonding, metal–ligand coordination, and guest–host interaction. Adapted from references [47], [51], [53], [56], [58], [59], and [60].

shows how hydrogen bonded brush polymers are able to self-assemble into a two-phase morphology, in which a polystyrene backbone acts as the hard domain in which soft brushes are able to self-assemble, *via* hydrogen bonding, into soft domains and behave like thermoplastic elastomers.[52] These materials are able to self-repair by bringing two cut ends together at room temperature, without external heat, and forming hydrogen bonds between C=O groups and amide or amine-functionalized ends. In a thermodynamic sense, self-healing temperatures above T_g will ensure sufficient chain mobility to achieve $\Delta S > 0$. Since bond reforming using hydrogen bonding interactions is an exothermic process, an overall $\Delta G < 0$ will facilitate self-healing at room temperature without external stimuli (E_{ext}).

In a manner similar to hydrogen bonding, metal–ligand supramolecular interactions are effective in the design of supramolecular healable polymers.[53,54] As a result of their optical and photophysical properties, metal complexes offer many advantages, and the reversibility of reactions can be tuned by different metal ion and ligand substitutes. Figure 5.8D shows phase separated supramolecular polymers based on a macromonomer comprising a hydrophobic poly(ethylene-co-butylene) core, with 2,6-bis(1′ methylbenzimidazolyl)pyridine ligands at the termini, assembled by complexation with Zn(II).[55] Upon exposure to UV light, the metal–ligand motifs are electronically excited and the absorbed energy is converted into heat. At the same time, elevated temperatures cause disengagement of metal–ligand bonds, resulting in the decrease of molecular weight and viscosity. Facilitated by the high temperature produced by UV radiation, a favorable ΔS is obtained such that lower molecular weight chains are able to diffuse into each other, reforming metal–ligand complexation upon cooling. Recent studies also took advantage of the coordination between Fe and catechol ligands, as well as boronate–catechol complexation, which resulted in a pH-induced, cross-linked, self-healing polymer with near-covalent elastic moduli.[56–58] As depicted in Figure 5.8E and 5.8F, the affinity between the metal and catechol is pH dependent. Another supramolecular assembly approach employed the host–guest interaction between cyclodextrin (CD) and ferrocene, and formed a transparent supramolecular hydrogel by mixing poly(acrylic acid) (pAA) possessing β-CD as a host polymer with pAA possessing ferrocene as guest polymer.[59] Redox stimuli could induce a sol–gel transition, as well as controlling the self-healing property, as shown in Figure 5.8G.

5.4 Summary

In summary, although the field of stimuli-responsive materials in general, and self-healing in particular, is relatively new, recent scientific studies have led to the developments of new technological advances in this fascinating and continuously evolving field. Although the complexities of transient physico-chemical processes represents a significant challenge, the future long-term payoffs will be substantial and will lead to a new generation of sustainable and environmentally compliant materials that will maintain their functions and appearance for extended times.

References

1. M. W. Urban, *Nature Chem.*, 2012, **4**, 80.
2. D. Y. Wu, S. Meure and D. Soloman, *Prog. Polym. Sci.*, 2008, **33**, 479.
3. E. B. Murphy and F. Wudl, *Prog. Polym. Sci.*, 2010, **35**, 223.
4. R. P. Wool and K. M. O'Connor, *J. Appl. Phys.*, 1981, **52**, 5953.
5. P. G. de Gennes, *J. Chem. Phys.*, 1971, **55**, 572.
6. Y. H. Kim and R. P. Wool, *Macromolecules*, 1983, **16**, 1115.
7. P. J. Flory, *Proc. R. Soc. Lond. A*, 1956, **234**, 60.
8. F. Liu and M. W. Urban, *Prog. Polym. Sci.*, 2010, **35**, 3.
9. K. Takeda, H. Unno and M. Zhang, *J. Appl. Polym. Sci.*, 2004, **93**, 920.
10. A. S. Hay, *Adv. Polym. Sci.*, 1967, **4**, 496.
11. D. M. White and H. J. Klopfer, *J. Polym. Sci.: Part A*, 1970, **8**, 1427.
12. K. Tekeda, H. Unno and M. Zhang, *J. Appl. Polym. Sci.*, 2004, **93**, 920.
13. C. M. Chung, Y. S. Roh, S. Y. Cho and J. G. Kim, *Chem. Mater.*, 2004, **16**, 3982.
14. J. Paczkowski, in *Polymeric Materials Encyclopedia*, ed. J. C. Salamone, CRC Press, Boca Raton, 1996, p 5142.
15. V. Ramamurthy and K. Venkatesan, *Chem. Rev.*, 1987, **87**, 433.
16. P. L. Egerton, E. M. Hyde, J. Trigg, A. Payne, P. Beynon, M. V. Mijovic and A. Reiser, *J. Am. Chem. Soc.*, 1981, **103**, 3859.
17. M. Hasegawa, T. Katsumata, Y. Ito, K. Saigo and Y. Litaka, *Macromolecules*, 1988, **21**, 3134.
18. M. Stevens and A. Jenkins, *J. Polym. Sci.*, 1979, **17**, 3675.
19. X. Chen, M. A. Dam, K. Ono, A. Mal, H. Shen, S. R. Nutt, K. Sheran and F. Wudl, *Science*, 2002, **295**, 1698.
20. Y. L. Liu, C. Y. Hsieh and Y. W. Chen, *Polymer*, 2006, **47**, 2581.
21. Y. L. Liu, C. Y. Hseih and Y. W. Chen, *Macromol. Chem. Phys.*, 2007, **208**, 224.
22. P. G. de Gennes, *Hebd. Seances Acad. Sci.,Ser. B*, 1980, **291**, 219.
23. P. P. Wang, S. Lee and J. P. Harmon, *J. Polym. Sci., Part B: Polym. Phys.*, 1994, **32**, 1217.
24. C. B. Lin, S. Lee and K. S. Liu, *Polym. Eng. Sci.*, 1990, **30**, 1399.
25. C. C. Corten and M. W. Urban, *Adv. Mater.*, 2009, **21**, 5011.
26. M. Zako and N. J. Takano, *J. Intell. Mater. Syst. Struct.*, 1999, **10**, 836.
27. M. Motuku, U. K. Vaidya and G. M. Janowski, *Smart Mater. Struct.*, 1999, **8**, 623.
28. R. Varley, in *Self-Healing Materials: An Alternative Approach to 20 Centuries of Materials Science*, ed. S. van der Zwaag, Springer, New York, 2007, p. 95.
29. S. J. Kalista and T. C. Ward, *J. R. Soc. Interface*, 2007, **4**, 405.
30. S. J. Kalista, T. C. Ward and Z. Oyetunji, *Mech. Adv. Mater. Struct.*, 2007, **14**, 391.
31. D. Ramachandran, F. Liu and M. W. Urban, *RSC Adv.*, 2012, **2**, 135.
32. B. Ghosh and M. W. Urban, *Science*, 2009, **323**, 1458.
33. B. Ghosh and M. W. Urban, *J. Mater. Chem.*, 2011, **21**, 14473.

34. B. Ghosh and M. W. Urban, *J. Mater. Chem.*, 2012, **22**, 16104.
35. R. Nicolay, J. Kamada, A. Van Wassen and K. Matyjaszewski, *Macromolecules*, 2010, **43**, 1191.
36. Y. Amamoto, J. Kamada, H. Otsuka, A. Atahara and K. Matyjaszewski, *Angew. Chem. Int. Ed.*, 2011, **50**, 1660.
37. Y. Amamoto, H. Otsuka, A. Takahara and K. Matyjaszewski, *Adv. Mater.*, 2012, **24**, 3975.
38. J. Kamada, K. Koynow, C. Corten, A. Juhari, J. A. Yoon, M. W. Urban, A. C. Balazs and K. Matyjaszewski, *Macromolecules*, 2010, **43**, 4133.
39. J. A. Yoon, J. Kamada, K. Koynov, J. Mohin, R. Nicolay, Y. Zhang, A. C. Balazs, T. Kowalewski and K. Matyjaszewski, *Macromolecules*, 2012, **45**, 142.
40. J. Canadell, H. Groossens and B. Klumperman, *Macromolecules*, 2011, **44**, 2536.
41. K. Imato, M. Nishihara, T. Kanehara, Y. Amamoto, A. Takahara and H. Otsuka, *Angew. Chem. Int. Ed.*, 2012, **51**, 1138.
42. H.-G. Korth, *Angew. Chem.*, 2007, **119**, 5368.
43. R. C. Osthoff, A. M. Bueche and W. T. Grubb, *J. Am. Chem. Soc.*, 1954, **76**, 4659.
44. S. W. Kantor, W. T. Grubb and R. C. Osthoff, *J. Am. Chem. Soc.*, 1954, **76**, 5190.
45. P. Zheng and T. J. McCarthy, *J. Am. Chem. Soc.*, 2012, **134**, 2024.
46. R. P. Sijbesma, F. H. Bejer, L. Brunsveld, B. J. B. Fomer, J. H. K. Hirschberg, R. F. M. Lange, J. K. L. Lowe and E. W. Mejer, *Science*, 1997, **278**, 1601.
47. F. H. Beijer, H. Kooijman, A. L. Spek, R. P. Sijbesma and E. W. Meijer, *Angew. Chem. Int. Ed.*, 1998, **37**, 75.
48. S. H. M. Sontjens, R. P. Sijbesma, M. H. P. van Genderson and E. W. Meijer, *J. Am. Chem. Soc.*, 2000, **122**, 7487.
49. J. R. Fredericks and A. D. Hamilton, in *Comprehensive Supramolecular Chemistry*, ed. J.-M. Lehn, Pergamon Press, New York, 1996.
50. P. Cordier, F. Tournilhac, C. Soulié-Ziakovic and L. Leibler, *Nature*, 2008, **451**, 977.
51. A. Phadke, C. Zhang, B. Arman, C.-C. Hsu, R. A. Mashelkar, A. K. Lele, M. J. Tauber, G. Arya and S. Varghese, *Proc. Natl. Acad. Sci. U. S. A.*, 2012, **109**, 4383.
52. Y. Chen, A. M. Kushner, G. A. Williams and Z. Guan, *Nature Chem.*, 2012, **4**, 367.
53. U. S. Schubert, C. Eschbauner, O. Hien and P. R. Andres, *Tetrahedron Lett.*, 2001, **42**, 4705.
54. F. R. Kersey, D. M. Loveless and S. L. Craig, *J. R. Soc Interface*, 2007, **4**, 373.
55. M. Burnworth, L. Tang, J. R. Kumpfer, A. J. Duncan, F. L. Beyer, G. L. Fiore, S. J. Rowan and C. Weder, *Nature*, 2011, **472**, 334.
56. B. Lee, J. L. Dalsin and P. B. Messersmith, *Biomacromolecules*, 2002, **3**, 1038.

57. N. Holten-Anderson, M. J. Harrison, H. Brikedal, B. P. Lee, P. B. Messersmith, K. Y. Lee and J. H. Waite, *Proc. Natl. Acad. Sci. U. S. A.*, 2011, **108**, 2651.
58. L. He, D. E. Fullenkamp, J. G. Rivera and P. B. Messersmith, *Chem. Commun.*, 2011, **47**, 7497.
59. M. Nakahata, Y. Takashima, H. Yamaguchi and A. Harada, *Nat. Comm.*, 2: 511.

Subject Index

Page numbers in *italics* refer to information in tables or figures.